Educated at Emmanuel College, Cambridge, J. D. Bernal did research at the Davy Faraday Laboratory and at Cambridge before joining Birkbeck College, University of London, as Professor of Physics in 1937. In that same year he was elected Fellow of the Royal Society, and was later honoured with the Royal Medal of the Society (1945) and the Lenin Peace Prize (1953). In 1963 he was appointed to the Chair of Crystallography at Birkbeck College, and he was Professor Emeritus there at the time of his death in September 1971. He was the author of several books, including *The Origin of Life* and *Science in History*.

The

J. D. Bernal

Extension of Man

A History of Physics before 1900

Paladin

Granada Publishing Limited
Published in 1973 by Paladin
Frogmore, St Albans, Herts AL2 2NF

First published in Great Britain by Weidenfeld & Nicolson 1972
Copyright © The Estate of J. D. Bernal 1972
Made and printed in Great Britain by
Fletcher & Son Ltd, Norwich
Set in 11/12 pt Baskerville, Series 169

Contents

The Leyden Jar
Benjamin Franklin
The lightning conductor
Coulomb and the law of attraction
Galvani: animal electricity
Volta: the electric battery
The Royal Institution
The lag in electrical development

Acknowledgements

The author and publishers would like to thank the following for the use of illustrations in this book: the Trustees of the British Museum for figures 6, 13, 28, 34, 35, 37, 42, 43, 51, 55, 57, 60, 78, 80–86, 88–95, 99, 100, 105; the Science Museum, London, for figures 9, 20, 44, 45, 47, 50, 52, 53, 56, 69, 70, 71, 72, 73, 75, 99, 103, 104, 110; *Scientific American* for figure 2; the University of London Library for figures 29, 30, 31, 38, 63, 66, 79, 96, 98, 106, 109; the Ronan Picture Library for figures 48, 58, 59, 97, 108, 111; the Royal Society for figures 68, 74; Cambridge University Library for figures 49, 66; the Royal Astronomical Society for figure 58; the Powell Cotton Museum, Birchington, Kent for figure 10; the Cambridge University Press and Professor J. Needham for figure 36; and Hauptamt für Hochbauwesen Nürnberg for figures 32 and 33; Verlag Karl Alber GmbH for figure 61, from F. Klemm, *A History of Western Technology*. Figures 9, 73, 99, 104 and 110 are Crown Copyright.

The author also wishes to thank: Mrs Nina Middle, who typed from tape recordings this series of lectures, a considerable task of interpretation; Dr Peter Trent for providing data and selecting, with Mr Stan Ward, illustrations from the slides used in the lectures; Miss Anita Rimel, who, because of his illness, completed the editing of the typescript for the author's approval; above all, Mr Francis Aprahamian without whose care in the final stages this book would never have been completed for publication.

Preface

This book is an attempt to record in my own words a set of lectures given in the Michaelmas term to first-year students in physics at Birkbeck College, on the history and nature of experimental physics. It aims to take physics up to the end of the period of classical physics, the end of the nineteenth century, just before the new discoveries of the atom and relativity were made. This marked a change as great as, if not greater than, that which occurred in the Renaissance and is called the first Scientific Revolution.

In view of the prestige of the new discoveries, I felt it was likely that the older parts of the subject would be largely forgotten and that something should be done to keep them in the minds of young physicists as a guide to the way physics has grown as a result of the interplay of necessities of navigation and gunnery, on the one hand, and the development of philosophical and religious ideas on the other.

So much of modern physics is now deemed to be theoretical and to be achieved by complex mathematical techniques in which the role of the experiment is necessarily played down. This is a paradoxical situation, for at the present moment physical experimentation is more costly and elaborate than it has ever been in the past. I did not continue the lectures into this modern period, not because I was uninterested in it, but because I knew it was being dealt with in detail by the other lecturers in the Department of Physics in the College.

Necessarily the historical element entered, but in a different way, because the men responsible for the great twentieth-century changes – Rutherford, Bohr, Planck, Einstein – are only recently dead and are treated as contemporaries, so that there is no danger of their work being forgotten.

In presenting a course of lectures verbatim, or approximately so, it is impossible to give the resulting whole the same finished appearance as would be achieved in composing a written book. Apart from the curiosities of language in going from the spoken to the written word, much simplification had to be made in the historical narrative and this has been presented without qualifications, but on the basis of historical studies in which I have been long involved.*

I have tried to make it clear that the opinions of the men of the distant past have a greater relevance than we usually realise. For instance, the idea of 'atom' not only retains the language of the Greek, Democritus, who first postulated it, but there is also an absolutely unbroken connection between the atom of the Greeks and that of the modern physicist. Moreover, the ideas of the Greek philosophers, Plato and Aristotle, fanciful as many of them are, also run through the whole course of development of physics from that time to this. Much of the progress of physics has consisted in the criticism and overthrow of these ideas, so that even when wrong they have played an important part in promoting later knowledge.

Not only the ideas but also the techniques of physics have ancient roots. The experimental method itself is a human construction inseparable from those of practical utility. Such developments as those of the cannon or the steam engine are both products of the piston water-pumps of antiquity, whether used for irrigation or for blowing organs. Originally, physics was a codification of the achievement of technologists. In the modern age the process is reversed: much of technology has become applied physics. Even such fundamental concepts as work and energy come from the field of practice. Even if the steam engine was rightly called a philosophical engine, depending on the understanding of the vacuum, it was at the same time made possible only by practical ironmongers such as Newcomen of Dartmouth.

I have thought that the historical method would be a very suitable introduction to the fundamental concepts of physics. I hope that those who read this book will be able through it to see something of the interplay between the theoretical and practical aspects of the subject.

J. D. BERNAL

*See my books *Science in History* and *Science and Industry in the Nineteenth Century*.

Introduction:
What is Physics?

These chapters are based on lectures given as a short intro-
ductory course aimed at explaining what experimental physics
is and how it has come to be like that. I have been giving these
lectures because I believe that it is useful to have some idea,
not only of what people believe now – which is, after all, only
a temporary stage in the development of physics – but also how
we came to think in that way and how the whole of current
physics is tied up with its history. The very language of physics
has grown up with the history of physics. This is linked, on the
one side, with the development of practical devices and, on the
other, with the development of human thought in the philosophical field.

We are too apt to think of science as something which exists
as a group of bits of knowledge. It represents what we know;
what we can use directly in practice and, further, is a basis for
finding out other things. At present science is all that and will
be, I imagine, increasingly so in the future; but science is much
more. It appears as a very rapidly moving river and the most
you can hope to learn now is, as it were, the course of the last
part of the river together with some general clues as to where
it is likely to flow in the future. You cannot predict what will
happen but you have to be ready to meet it, and that is why
the teaching of science, particularly the teaching of physics,
has to take this historical aspect very much into account.

It is reckoned, by actual analysis of references, that the
half-life of an average paper in physics is two and a half years.
That is, the chances are less than even of it being referred to
after such an interval. If this were strictly so, of course, any
benefit you received from a physics course would have vanished

by the time you reached the end of it. But it is not quite as bad as that, for the papers refer to new knowledge and even if the still newer knowledge catches up with that new knowledge, most of the old knowledge still remains satisfactory if it is properly interpreted.

For instance, we still talk about atoms, but we would not like to say any longer, as did the Greeks or the Victorians, that an atom was eternal and unchangeable. The fact that an atom is not eternal and can change suddenly and violently is, nowadays, unfortunately, only too familiar to us, but it is still very useful to have atoms to talk about. In the same way people could talk about men or cats before there was any anatomy or any knowledge of evolution. All the terms of the past and the methods of thinking about them are still useful. Most of what students do on the mathematical side of physics is to be found in Newton's *Principia*. The basic idea of dynamics, the equations of motion, the ideas of momentum, of inertia, of mass and of acceleration, which were applied by Newton to large bodies like the earth or the moon, or to moderate-sized bodies like cannon-balls, turned out to be very useful for handling minute hypothetical entities like atoms or even smaller things like the fundamental particles that build them, or even things that we would say are not material at all like the particles of light that we call photons. So the past still lives in physics and it is worth seeing it in its contemporary pattern at the time when you are learning to use it.

The separation of the sciences

Now, with that slight introduction, I want to go straight on to considering physics as such, as different from other branches of science. How do we now, for convenience, distinguish physics from chemistry or mathematics? It used to be a much easier distinction to make. In fact, when modern science was growing up, from the time of Galileo to the time of Newton, all the sciences were very much joined together. A single man like Hooke could do absolutely first-class research in pure mathematics, in physics, in chemistry and even in biology. Towards the end of that time the sciences were just beginning to separate and after that they continued to separate more and more. By the nineteenth century, in fact up to the time I was in

Cambridge nearly forty years ago, the science faculties were almost completely separate – but there were already signs that they were coming together again.

Just at this moment we are witnessing a great convergence of all the sciences. In order to be a good biologist now, for instance, you have to know not only what might be called school physics, but quite a lot of really modern physics – quantum physics and so forth – and a great deal of chemistry as a basis for biochemistry. Conversely, the physicist himself, even if biology is not part of his subject, is obliged to know something of it because he may find a great deal of his work will be concerned with biophysics. The problem for a physicist, rather than for physics as a subject, arises because physics is increasingly penetrating all the other parts of science. This is already evident in the names of the new hybrid subjects. We have long had one called physical chemistry; now we have a subject called chemical physics which is different, not so much in the proportion of physics and chemistry that come into it, as in its central interest of helping chemistry in the first case and of extending the range of physics in the second.

Now we also have biophysics and biochemistry. It would appear that physics is spreading towards biology on one side, while on the other the mathematical aspect of physics is also becoming much more marked, especially now that we have a growing symbiosis between physics and mathematics in computers. For the computer contains both the purely physical element in its actual components and mathematical logic in their arrangement. To develop and make the transistors or magnetic memory elements or the newer devices that will replace them, needs a great deal of physics, but although the connections between the computers are actually made with material wires or printed circuits, their lay-out and set-up are really pure mathematics.

The nature of physics

What I am saying first of all then, before I try to define physics, is how difficult it is to define. Nevertheless, we can always fall back on this – and it is a very convenient thing to say: *physics* is a limited subject. For examination purposes physics is heat, light, sound, electricity and magnetism, with a small allowance of atomic physics; here we will look more closely at the 'examination purposes' and at how such physics came into being.

The reason why we can block out this particular part of knowledge and experience and call it *physics* is because it deals primarily with what might be called *the extension of the human sensory-motor arrangement*. A man may be deprived of everything, deprived of tools or clothes – in the definition of the Yahgans of Tierra del Fuego for a poor man, he is 'body only' – yet he possesses a most complete set of *physical apparatus* in the shape of sense organs to register the external world and muscular effectors to change it. It is true that he also possesses an even more elaborate *chemical apparatus* for digestion and the maintenance of metabolism. But while the latter became intelligible very recently, and then only partially, by sophisticated biochemistry, the physical sensory-muscular apparatus is comparatively understandable, or at least it was there that man's rational understanding of his world which we call *physics* began.

To combine sensory and motor experience, we must begin with sight. We use the word *see* very much as a form of understanding; you say 'I see it', meaning that you see the connection. Actually, sight comes very late in animal development. The earliest animals did not see very much; they worked largely on smell, which is a much more delicate, much more sensitive, much more varied way of sensing the environment, but it is essentially chemical and not physical. Now you cannot argue about smells: you cannot deduce anything from them at all; it is all subtle discrimination, memory and association. But you can argue about sight: you can say this is near or that is far, you can say this is on one side and not on the other side, you can relate sight to motion. One of the very big achievements in practical physics, yet a definitely pre-human one, is throwing something, especially throwing something at a mark and hitting it.

Another very similar sort of achievement is projecting yourself in a free jump. Early animals, for a very long time, right back to their first coming out of the sea, managed to walk just by putting one leg in front of another and had built-in mechanisms for doing so in order. But jumping is a different matter. Fleas or sand-hoppers just jump and it does not matter very much where they come down; but to jump like a monkey that can take a twenty-foot jump and grasp a branch and hang on to it, needs a very much higher degree of eye-hand co-ordination, which is really the physiological basis of physics. A further stage in really human evolution came when a man not only threw a stick around but used it. He used it for poking something, for digging something up or bashing something. The use of a piece of wood or a stone is first of all an extension of the human frame, limbs and sense organs together. The naked man with his hands and teeth can do a certain amount: armed with a stick and a stone he can do far more.

Human sight

The physical extension of man's awareness and activity came first on the muscular effector side rather than on the sensory side – mirrors and lenses came far later than tools. Nevertheless, for understanding the world, the senses, even unaided, could go very far. This is especially true of sight. This is because, in particular, human sight is as much in the brain as in the eye. In the course of evolution the brain came to have a built-in analyser of the image presented on the retina of the eye. It is this analyser that turns mere variegated and moving patches of colour into a world of objects that can be distinguished and manipulated.

In fact, we see things dynamically rather than statically: if, by an ingenious device, the eye can be kept fixed on any particular object, it cannot see it at all. The eye cannot detect light, the eye can only detect changes of illumination; the eye is always wandering around, as it were monitoring the experience of the outside world. Seeing is not a passive perception – it is an act. The Ancients had indeed always thought it was an act; for a long time they considered that you saw things by projecting something out of your eye, an eye beam; it catches the thing and brings it back to you and that is how you see it. Consequently, of course, your eye becomes a very powerful magical instrument – if it is the wrong kind of eye it can do all kinds of damage to the person it looks at. These were, in a sense, rationalisations of what we now know as the *act* of seeing; and that leads immediately to the question of movement.

Muscular sense

Muscular movement is part of our sensation, part of the action. We have always talked about the five senses but there are more like seven senses and one of the senses is the muscular sense itself. We have, built into our muscles, a very complicated set of strain gauges to measure the degree of pull they are exerting. Inside the muscle there are certain little organs called spindle fibres that have their own nerve quite separate from the one that works the muscle, and this tells us how tight the muscle is at any given time. For even when we are not moving our limbs, we are sending messages to all our muscles; the limbs are very carefully balanced so that the muscles, which are arranged in pairs – like the flexors and extensors – are exactly and actively balanced in what is called tone. That is how you can gauge, for instance, without moving at all, the weight of something in your hand. More orders have to be given to the muscles that are pulling up than to those that are slackening off, so as to keep the hand steady, and the difference of sensation in the spindle fibres of these two sets of fibres gives the weight. The old mechanics of the 'five powers' was only a further sophistication of the body's ways of lifting things by lever, pulley, windlass, plane or screw. We shall see later how, by the use of these simple and extremely useful devices, a quantitative mechanics could be built up – man-powered in the first place. The first step towards this is the study of the lever as a balance. Balance in itself is built in to the human frame. We need it more than most animals because we walk top heavy on two feet and have to learn in infancy how to keep our balance, for which we are furnished with a levelling compass in our ears.

Force

The whole idea of equilibrium is really derived from levering operations. It was much later, when the mechanics of the whole body was studied, that the outer part of the body came to be understood as a system of levers. This analogy is important and we will discuss it when we come to the history of the concept of force in physics (p. 213). The natural starting point is the fact that *force is directly felt*. It is felt by the muscle spindle mechanism just described. Now, students of physics have been taught that force is a secondary idea. Since Newton, we look on force as rate of change of momentum or mass acceleration. But the primitive view has some justification. Force is as directly felt as things are seen.

Then there is the concept of inertia; the mathematical idea of inertia came very late, as we shall see (p. 214), but the feeling of inertia is also an immediate one. We learn the difficulty of setting objects in motion by experience. Objects may look the same, but if you try to move them about you realise they may be very different indeed. There is the old trick of painting bits of cardboard as heavy lead weights so that the person trying to lift them falls over backwards.

Hearing

Hearing, in essence, is another mechanical sense, but the time element is the principal aspect. The sound vibrations are not transmitted directly to the brain but pass through what is effectively a harmonic analyser which we carry inside our ears, each receptor resonating to a definite frequency, so that we hear notes rather than sections of a complete sound track. This is a very fundamental idea which students will find later in physics, the two ways of looking at any disturbance, regular or irregular: the direct way, the actual displacement tone graph, and the indirect way of analysing it into frequencies. We do this analysing into frequencies for ourselves in music; following the pre-human stage of interpreting cries and noises, we recognise them by their harmonic analysis, and that is why it is called 'harmonic analysis', because it all arises from the musical concept of harmony.

Touch

Then we have two other sensations that are vital to physics which come in in a different way; one of them is touch. Now touch is still somewhat mechanical, but it is a very, very complicated subject and its elucidation has only come about in the last thirty years or so. Students learn, and have possibly already learnt at school, one of the neatest laws of physics which is Amontons' law of friction. It has been taught since

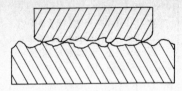

1   Magnified cross-section view of a block resting on a slide, showing actual area of contact

around 1670 – a very practical thing too – and no one realised until the 1930s that it was complete nonsense. The force, you learn, needed to move an object along a rough surface, depends on the nature of the surfaces in contact and on the forces pressing them together, but it is completely independent of the area of the actual surface of contact. Now, if any student said that this must be nonsense, he would probably be told that this is a law of science, that it has been proved to be correct for nearly three hundred years and that one must not ask awkward questions. In fact, as my sophisticated readers have probably learnt, though the law is perfectly true, the explanation is nonsense. In fact, the friction is not independent of the area of contact, it is strictly proportional to it, but the area of apparent contact is not the same as the area of real contact. Look carefully at a block resting on a slide (see Fig. 1). The surface is in contact but only in a few small places. If you put more weight on the block it settles down and meets the slide in more places. The area of contact can be measured by a simple process of covering the surface with paint and so finding out which parts are touching and which are not. But no one thought of doing that.

The kind of thing I mean is what you feel when you touch anything very lightly, when you are apparently only just touching a surface so that you can feel it. The French call it *effleurer* – just like stroking a flower; you are certainly pressing on the surface in very few places with about 1,000 atmospheres pressure at the point of contact which is enough to break down the surface molecules of the skin. As people now realise from the damage done by stiletto heels, you can get very considerable pressure simply by lowering the area of contact. Touch is a very delicate sense. It depends on two sets of receptors. One set are the very sensitive receptors concerned with the actual touching, which are more or less limited to the fingertips or the lips – these are special receptors that are very good surface measurers. It is only in the last few years that we have been able to get any kind of gauges nearly as good as the feel of the fingertips. The miller, who just feels the flour between his finger and thumb, has a better gauge of the quality of the flour than any scientific method we can find to do this for him.

Besides these, there are a deeper set of touch receptors able to appreciate the yield of the surface pressed on, so as to tell whether it is hard or soft in the same way as the first set tells us whether it is rough or smooth.

The sensation of temperature

Finally, there is a type of detector of an entirely different kind – temperature or, rather, heat-flow detectors. As you all know, heat is simply a mode of motion consisting of vibrations; but nothing so slow-moving as nerve circuits could analyse those vibrations. What we call a direct sensation of heat or cold is one, not of temperature, but of change of temperature. You probably know the trick of putting one hand into a bowl of hot water and the other into one of cold. A bowl of tepid water will then feel cold to the first hand and hot to the second. What is being felt here is really the rate at which heat flows into or out of the skin, which is mainly controlled by the heat conduction of what it touches. Metals feel either very hot or very cold. Wool is comfortable at all temperatures.

The sensation of temperature is only physical in origin: what is actually detected is a chemical change analogous to the two purely chemical senses – smell and taste – about which I will say no more here except that they are the most sensitive of all.

Taking together the senses and the voluntary motions that go with them, I come back to the simplest definition of physics as the subject that deals with the things you feel and the things you do. But that is what might be called 'real' physics and much of it can be learned without benefit of any book knowledge at all. Some of the basic physical devices I will discuss were developed by completely illiterate and even socially very backward people. The amount of physics you need to be a fully functioning Eskimo is very considerable and many things we do not understand yet are used by Eskimos for various purposes. One that I came across, for instance, was in connection with some war-time experiments on making reinforced ice. The Eskimos knew that by taking moss, damping it and letting it freeze, they got a material which was so hard and tough they could use it for long-wearing sleigh runners. That is something which we with our advanced physics would find it hard to predict. That kind of physics, the physics of knowing how to deal with the world by an ingenious combination of materials and shapes, is something that must be discovered by practice.

Social aspects of physics

Behind that, however, there is another kind of physics that has played a much larger part in history. Technical developments on the practical side will get you so far and no further. A more sophisticated kind of physics came from the attempt to understand – and put into words in some way – what these practically useful experiences were. This aspect is almost purely social. The problem of how to organise human relations with the universe as a social act – as, for instance, with the movements of sun, moon and stars and the farmers' calendar – gave rise in history to the main lines of physical theory and, in the long run, even paid off in practice. This was, as it were, an attempt to find a language for describing the external universe that could be really useful. A great deal of the history of physics is the history of the refining of that particular kind of scientific language, for ordinary language is full of imprecisions and errors. V. Gordon Childe* writes that language is, of all elements of culture, the most persistent and hard to change. All our present-day languages, though mixed, cannot be very different from what they were in the Old Stone Age. The basic words and their connections were originally made for Old Stone Age conditions and yet we have to use them as time goes on for everything – the swords, cannons, spacecraft, masers and so forth – that we keep on inventing. The Old Stone Age languages are always being modified but never recast. The main basis of the language and the main basis of the thought behind it still remains a fossil of old thinking and we will have to go on a long way before we can think

*Society and Knowledge, Harpers, New York, 1956, p. 94.

straight enough to get away from it. For instance, some notions which we now consider as purely mathematical, such as the right angle and the square, are technical and social in their origin.

The four-square gridding that we call the Cartesian co-ordinates goes back a long way, but the notion of the right-angle cross is not mathematical or physical in its origin. The square Roman camps, for instance, were laid out precisely in that way: there were two main streets with the command post at the crossing. The Roman camp was originally modelled on a village: a village with a great number of people on the move, but a planned village, nevertheless, of the eighth century BC in Italy. But villages had been laid out in such a four-square plan since the Stone Age for purely social reasons, corresponding to a four-fold division of the tribe into four non-intermarrying sub-clans. Each of these had its appropriate sign and totem animals.

The idea of the correct lay-out for the settlement is the idea of law, of rule. Rule itself is a social word which was afterwards given a mathematical meaning. The same is true of the points of the compass, corresponding to the orientation of the camp: a transition from the primitive social arrangement to a highly sophisticated geometrical one.

The four elements

With these go the four elements – earth, water, air and fire. We are not so interested in the four elements now, we only talk about them metaphorically: the conflict of the elements, or to be in one's element. To us, scientifically, elements are very specific substances consisting of a very particular kind of atom. The Greeks had four very simple elements – *fire*, that was said to be above the air; *air* itself; *water*; and *earth*. Before that they were even less connected with nature; the Chinese, for instance, included metal and wood as elements, practical elements. These were all attempts to picture the universe, particularly those parts of the universe that we cannot alter. It took some time to realise that there were some things you could do – you could catch animals, you could cook food, you could plough the ground and, later on, sow and reap. But there were some things you could not do – there were the sun, the moon and the stars beyond your reach. You could not do anything about them, but they could do a lot of things to you; they could warm you or they could let you freeze, they could blow winds on you or they could bring rain or they could refuse to bring rain. These were elements of a world that had to be observed but that could not be manipulated in any certain way, though something could be hoped for from prayers and magic.

The observation of the heavens

The knowledge of the observable, the motion of the heavenly bodies, was to be the clue in a kind of long detective story through which the ideas of physics were to be derived. The first, naïve visions of the world plan revealed that it was a very homely kind of world in which there were four walls and a roof with various lights in it that kept moving about – two big lights and many small lights. That was quite a good description but not much use by itself. What men needed to know was when the lights were to be turned on and off, when the rainy weather was going to come, when the river was going to rise. All these things required sustained observations and led to the discovery, the most fundamental discovery of all, that there are things which are *regular*. Now you might say that for this you do not need elaborate aids: there is a day, and a day is always the same length, but how do you know it is the same length? Unless you have a clock, there is no way of knowing it is the same length. A day does not always feel the same length – some days feel long and very boring and some days feel exciting and very short. Further, if you count the days after a full moon you will find that there is another full moon some 29 days later.

Counting, of course, is very old indeed. Figure 2 shows the oldest scientific document we know of so far. It is a bone handle from Central Africa, about 12,000 years old, and it is scratched. I have tried to draw the scratches on one side of the handle and you can see whether you can make any sense out of them. On the other side there is a different set of scratches – 11, 13, 17 and 19. Well, clearly, the man who made this could not only count, he could multiply and divide because he had scratched the prime numbers on this piece of bone.

Much can be deduced about events in the heavens by just counting long enough. If you observe the moon for a whole

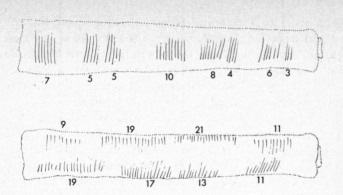

2 Scratches on a bone handle from Ishango in Central Africa, giving evidence of a knowledge of multiplication and of prime numbers some 12,000 years ago

month, beginning when it is a thin crescent at sunset, for nine days it gets bigger and for the next nine days it is nearly full, finally for nine days it gets smaller, after which you do not see the moon for one or two nights. No matter how long you count, you cannot get an even number of days in a month or months in the year. It is a shock to find that although at first the heavens seem to move regularly, there are no precise regularities; and the whole of mathematics has been built round the impossible task of getting the calendar right, or the exact number of days in a month or the number of months in a year. They are all irrational numbers and they are irregular in other ways as well. The number of days in the summer is different from the number in winter – that takes some explaining! Neither the Babylonians nor the Greeks could do it and it was not really explained until Newton's time.

Most people never bothered with these things. Primitive hunters never had to look beyond the next meal or at most the next full moon but, ever since agriculture began, the calendar has mattered and someone has had to look after it – magician, priest or royal astronomer. There was no need to have many of them but they were considered to be important, even essential, to the running of the state. Thus mathematics and astronomy came to be regarded as higher forms of knowledge – they were used for the original calendrical and observational purposes but they were also given purely magical purposes. The Great Pyramid in Egypt is a piece of what might be called embodied mathematical nonsense; it is extremely precisely oriented within a few minutes of arc, correct north and south, east and west. It is exactly square, the sides are sloped exactly right and the tunnel down the side points exactly to the North Pole. Through it the priests could see the North Star of the time – for it was a different North Star then. All this was an expression of sophisticated mathematics and astronomy in the service of extraordinarily primitive religious beliefs, about the incarnation of the Pharaoh as the sun god. But the fact remains that astronomy and mathematics have been top sciences ever since. The other aspects of physics to which I have referred – heat and light and so forth – were hardly admitted as sciences until much later. It was only in astronomy that it was possible to cope with the observables by devising mathematical tricks, starting with something very simple. Men had a complete computer set on their fingers and these could always be used for any mathematical purpose in the first place. This could afterwards be improved on by making the next type of computer, which is ten beads on a string. Then, without the fingers, you get the abacus; from the abacus came decimals and ultimately notation and from that the whole of mathematics. But I am not going to talk any more about that because it is the theoretical side of physics.

Measurement

It was by developing theory – for instance, by developing astronomy first for the calendar and secondly for navigation – that the idea of measurement was incorporated into science. It entered in two ways: to regulate customary exchanges, which turned into trade and taxation; and to regulate the movements of the heavens insofar as they did or were deemed to influence things on earth. Both come together in the organisation of a divinely ruled agricultural state like that of ancient Egypt.

The beginning of both is the *measure*. The next step in science after the counting of individual objects is the counting of repeated standardised objects like *units* of measure. Now, what is such a unit and what in the first place did it measure? Even today, if you ask for a measure in a shop and do not specify what you mean – such as a tape-measure – you will get a measure of volume. You would mean a basket of some standard dimensions or a jug, say, in which you could measure the two things that really matter to be measured – you measure grain or beer. The measure of grain was the basic *metron*, it was a beginning of all measurement. Length was next measured in paces, essential for the measurement of area of land. It was necessary to know how much of a field was needed to obtain so much grain. Think of a central organisation that requires a tax to be levied in so many baskets of grain; the tax collector has to be able to prove that the peasant can produce it, or the peasant has to prove that he cannot. This is the first stage of measuring the ground, in the language of the Greeks, *geometry*. Fields were measured for their yield; in China, for instance, the standard land measure is called the *mou*, but the old *mou* is not always the same area. It would be a large area where the land was poor and a smaller one where it was rich. Everywhere the *mou* was the amount of land that should yield the same number of baskets of rice.

The measure of *weight* came later for things which could not be reduced to lengths or volumes, beginning possibly with raw wool and going on to metal, especially precious metal. It depended on the invention of the balance (pp. 49f.). Measure was introduced everywhere for its utility function, devised *ad hoc* for every purpose and only after many centuries integrated into a system like the metric system.

The last thing to be measured accurately was the first to be measured at all – and that was *time*. When people became particular about time, they were not satisfied with counting days; they wanted sub-divisions and it required a certain amount of astronomy to get these equal. They had to set up posts to measure the sun's shadow and mark how it moved and devise the *gnomon* and the sundial in order to divide the days into twelve hours corresponding to the twelve months of the year. For astronomical purposes this could be divided into sixty smaller parts, the *minutes*, and each of these into sixty even smaller parts, the *second* minutes – and so on.

This arrangement by sixties is an example of astronomy, astronomy imposing itself on common sense. The first countings were digital, in tens. Then it was noticed that there were roughly 360 days in the year and so it was convenient to divide the circle of the heavens into 360 steps or degrees – a degree means a staircase step, a grade. Starting with astronomy, it was convenient to count other things in sixties. The Babylonians counted sixty minas to the shekel in their money tables. Their economic calculations were made on the basis of astronomical calculations for which they had already made elaborate tables. The upper and nobler half of mathematics was concerned with the heavens and the other half was concerned with buying and selling and satisfying tax demands on earth; but both led to the development – the purely technical development – of computational methods. We see this now more clearly than we did because it has happened to us in our own times. We are very pleased with the wonderful things electronic computers can do, but the people of 5,000 years ago who introduced the sexagesimal notation found the concept of place enabled them to carry out and tabulate such functions as roots and trigonometrical functions which went beyond what they could do on the abacus or on their fingers. For them it must have been as big a jump in physics and astronomy as the computer is for us today.

These remarks give in the barest outline some indication of the motives that led to the first understanding of what we now call the physical nature of the universe and what was long called natural philosophy. This was connected with astrology, which at the time was considered to be terribly important and, however nonsensical it may seem to us now, this was to turn out to be a very happy development. It is a curious thing that since the weather occurred regularly, although people's business affairs did not operate at all regularly, it was believed that they ought to and that if only they did happen regularly everything would be right. So you had to fit in with the big machine, you had to find out how the stars were working and so forth, and you had to find a correlation between the earth, the microcosm, and the macrocosm (the large outside world). This, essentially, was astrology. Most of the physics up to the seventeenth century was motivated by the ideas of astrology. Much of our work consists of getting rid of or transforming those ideas, and this is in part what I shall be talking about.

Experiments

Next I want to deal with what we call the physical method. The physical method as we know it begins with observation and goes on to experiment. What is experiment? An experiment, from its name, means an experience that is often referred to as a trial. People are obliged to experiment, to find out by trying things out, whether they will work. A considerable step forward was taken when it was realised that you did not necessarily have to find this out on the full scale all the time. In the early days, for instance, when people were already developing mining – or at least tunnels – they made quite good tunnels, mostly for conveying water. If you are driving a tunnel from two sides of a mountain, you need to know where you are, especially if you have to make a bend in the tunnel. What they did was this: they took strings and stretched them inside the tunnel: they then went out into the field and stretched the same strings at the same angles in the open. They were then able to see how they were going and which way they ought to turn to make the two pieces join up, down and sideways. This was all done at full scale and then some bright person thought that you could really reduce the scale to a tenth or so and do it on a piece of paper instead of having to do it in a large flat field.

In due course, this led to the idea of the plan. The Egyptians and the Babylonians (p. 57) made plans of towns, of fortifications, of fields and other operations. Somewhere or other – though I suppose they have been lost by now – there must have been plans of the pyramids. They must have been drawn out in dimension: we know that because here was a mathematical problem, a piece of mathematics. They desired to build a pyramid so many ells long and so many high; they had to calculate how many stores were required, how many jars of beer and how many loaves of bread to feed the men, supposing the men could do such and such amount of work a day. Those are the Egyptian elementary textbooks of 2000 BC. They had to work out – and, as you know, it takes a bit of thinking out even now – the formula for the volume of a pyramid. But all these things, all these approaches to the introduction of mathematics, were part of a way of doing big things on a small scale. The fundamental realisation was the mathematical realisation of what we call the principle of similarity. Of course, as a notion it is extremely old, because it was used not only for objects but for animals and people. In the Old Stone Age they drew pictures of animals, usually fairly full-scale, and then drew pictures of spears hitting them, and this was the principle of similarity. Later, they would draw the animals quite small: they would draw them and tattoo them on each other and it helped them in the hunt. Here the idea of a scale came in – scale, incidentally is another of these practical things: it means the steps, the ladder. When the Egyptians had to make one of these magnificent large decorations on temples, they drew up a Cartesian net, a piece of graph paper, and they set ordinary workmen to do the work on the large scale. The artist drew the design on a small piece of paper and gave the co-ordinates, and the workmen put it up at a scale thirty times as big.

They had the idea of scale, but experiment was the trial. Now the best example of trial is not in physics but in chemistry. If you had discovered a mine and found some ore minerals and wanted to know how valuable they were, you could, of course, take the whole pile of ore and extract the gold from it. You would then know how much gold was there, but you would rather like to know beforehand whether it was worth doing. So you took a little sample and assayed it in a small way. Instead of using a large balance able to weigh half a ton or so, you used a little balance able to weigh a fraction of a pennyweight. There, then, you have the idea that you can get everything to work by doing it in a small way: and by doing it in a small way you could see what to do in a big way.

The controlled experiment came a good deal later and some of the first and best experiments, which we will be discussing later, were those made by Galileo who discovered the law of falling bodies by experiment or, to be more accurate, what he really did was to discover which of the laws that had seemed to mathematicians to be equally good theoretically, was the one which crude nature actually used. But his experiment on falling bodies was, in a sense, a little too good. What I mean is that he said: 'I have carried out this experiment one hundred times and it always gives exactly the same result' – he had not been trained in statistics. We would now say that he must have been telling lies, but the point is that by these experiments you interrogated nature or, in the language of the times, 'you torture nature and force it to reveal its secrets'. In a literal sense you do alter things by forcing them to show what they will do under different circumstances and then, from that point of view, find out their essential nature.

The building of theories

This leads, as it were, to the final aspect of the experiment. Having got the laws out of the experiment, the very simple laws like Boyle's law or Hooke's law, which are straight experimental laws, you want to see what it all means, what is the interpretation, what is the theory of it, how you can really understand it. And this building of theories, of really good theories, logically ought to have come last, but in actual historical development it came first. The first thing was to know the answer, the last thing was to find out if it corresponded to nature: the general view was, if it did not correspond to nature, so much the worse for nature! Nature was irregular, could not be relied upon, but the ideal thing – and Plato was the chief promoter of this doctrine – was that there is astronomy which consists of perfectly spherical bodies moving in perfect circles. This is real noble astronomy: the actual bodies in the heavens do not quite behave in that way, but they ought to. You can, in fact, study ideal things best and real things are very poor and distorted images of them.

The working out of these distortions came later. For instance, Kepler, having conceived a rather beautiful idea as to how planets ought to move, actually measured how they did move and found that there was only a difference of eight minutes between the theory and the practice, which could easily be neglected. Most people had accepted the theory, but he was particular. He was sure about his measurements and he said this difference must be real. With that he discovered that planets did not move in circles, they moved in ellipses, though from looking at these ellipses you could not tell that they were not circles. That is how observation, by not fitting, leads to continual improvement of theory.

One of the first people I worked with in physics was Kapitza. He was a nasty kind of experimental physicist; he never went in much for theory, but he had, of course, all the best theoreticians around him in Cambridge. He would ask them a very simple question – what would be the electrical resistance of such-and-such a pure metal in such-and-such a magnetic field? Well, the theory of the conduction of metals in magnetic fields is well known, there is no need to work it out. But then Kapitza would do the measurement and find that the theory was wrong – well, not seriously wrong, but by a small factor. And then, of course, the mathematician would go back and look at the theoretical calculations and would find that he had used approximations – which all mathematical physicists do, that he had neglected terms of higher order, that he had done various things which on careful examination he found he should not have done. He then produced another answer which was now, of course, reasonably in agreement with the experiment. Then Kapitza would say – well now, supposing we make the magnetic field ten times as strong and here, of course, he was perfectly safe and protected because no one had made a magnetic field as strong as this, so the process was repeated. Thus, theory is continually refined and criticised and destroyed and reformed by experiment. If you look at modern physics, which I will not be dealing with in this book, you will find plenty of examples of this – for instance, with mesons and other new particles – where theory has thrown doubt on experiment. And there are plenty of other examples where experiment has proved the theory wrong. Of course, the basis of modern physics is the whole of the quantum theory which arose out of the limitations of the previous theories, of their not fitting the experimental evidence.

Modern physics

I have given you here a very general account and I only want to say a little more in relation to modern physics. I will not be dealing with modern physics for the very practical reason that it is modern and you will get most of the relevant history of modern physics elsewhere. When I say 'modern physics', I can be very precise: physics after 1896 when the first breakthrough was made and, a most unexpected thing, made experimentally and not theoretically, in Röntgen's discovery of X-rays. With the discovery of X-rays, the clues came which enabled a number of other anomalies to be cleared up. Let us call the date 1900 for a first approximation as it is a very convenient date.

Modern physics, what is called atomic physics, is quite different from, although it does not contradict, the old physics, as you will see from the study of its history as you go along. What I want to do in this series of lectures is to take you up to 1900. I remember when I was first an examiner in this University – a rather long time ago, in the year 1937 – one of the older examiners said when we were looking at the B.Sc. Degree question papers: 'There is a question here about the electron. You have no right to ask that question, it is not in the syllabus.' We looked up the syllabus and, sure enough, the electron was not in the syllabus. The syllabus had been made in 1896 and, although the electron had been named in 1891, it was not 'respectable' and it was not included.

Indeed, the kind of physics which existed before all the new discoveries was a very complete doctrine, the most complete doctrine imaginable, which is now essentially called descriptive physics, that is, it consisted of the measurement and interpretation of what are called the properties of matter. For instance, you are told what the specific weight of gold is; you are told the coefficient of expansion of such-and-such things; you are told the refractive indices and you are given a certain number of measured constants which manipulated suitably can enable you to discover other things you are going to measure, or to predict them. We were told very severely that that was the nature of physics: if you were to ask *what* things are, you could make an answer, while you could never answer questions of *why*. For instance, why has water got an anomalous structure? Why has water got a point of minimum density? That is a fact: you learn the fact and if you learn it well enough and put it into an examination answer, you will get full marks. But that physics consists of knowing the laws. For instance, knowing Boyle's Law is good enough for school, but it is not good enough for university. You have to know that you can correct Boyle's Law, you can introduce variables that will improve on Boyle's Law and fit the facts better. And the kind of descriptive physics which includes among its laws some which have a rather different status – a better status, like Newton's Laws of Motion – is the part of physics that covers most of the field.

The exception, where you begin to impinge on what I call explanatory physics and not descriptive physics, is where the atom or any hypothetical unit comes in. It really came into physics to some degree in the nineteenth century. Curiously enough, it had come into chemistry at the beginning of that century: it was never quite respectable, but the atomic theory in chemistry does date back to 1804. The atomic theory in physics, however, does not really date back, in any full sense, earlier than 1913 when Bohr's theory of the atom appeared. But I would point out that the concept of the atom was used to explain Boyle's Law and the kinetic theory of gases. It was used to a minor extent in electro-chemistry, in Faraday's unit of electric charge which is, as we know now, the *electron* but was considered to be a hypothetical unit. Now, you do not want to reduce things only to observable constants, you want to reduce the number of observable constants themselves. The object of physics at the present time – it may change again – is to find the underlying regularities; that is, not just to say that gold is yellow and silver is white, but to find out *why* gold is yellow and *why* silver is white. We do not know yet: we have no idea. Gold is a fairly complicated atom and we cannot as yet deduce from first principles why gold should be yellow. We have no reason to doubt that it could be done and possibly with a good computer it could be done now, but most of these facts are interesting, not so much in themselves, but more in what lies below them. For instance, we would very much like to know what metals and what alloys are super-conducting at what temperatures. The only way at the present moment to do that is to try to find out experimentally. If we had a proper theory we could do it mathematically.

These, then, are the generalised tasks of physics, but there is still an upper physics and a lower physics. I will be talking only about the lower, experimental, physics, about things which are detectable and measurable. The higher physics, the mathematical physics, still deals with a world that is slightly less perfect than Plato would have liked, but neverthe-less is for any given time the most perfect explanation of phenomena which you can get.

In this first lecture, I have summarised the field and will deal with it more specifically next time. I think everyone ought to look at least at one original textbook in physics, to see what physics was like at any particular time. And one of the nice things to remember is that the further back you go, the more pleasant physics is to read, because people were more human then in the way they wrote about it. In modern physics, I would only mention Dirac's classical book on *Quantum Mechanics*. It is not exactly a delight to read, but it is one of the best written of its kind. The older textbooks were written at a time when people did not know as much as we do and therefore there is no particular difficulty in our understanding their content.

2

Ancient Science

Table I attempts to illustrate the time scale of developments in the various fields of physics from about 4000 BC to near the end of the last century.

First we have mathematics, touched on in the last chapter, starting with aspects of simple counting and leading on to geometry. Then, closely related to that, was the study of astronomy, because a great deal of astronomy is just counting – counting days, months and years. But the section of the Table with which I shall deal in this chapter, will be the third column, dealing with the more practical aspects – mechanics, phenomena of rapidly moving bodies; then to inertia, which appears in dynamics; and so to heat, which is closely related to a key discovery of man, that of fire. Finally, there are two discoveries which came very late in the history of mankind, although their first manifestations were certainly known even in the Old Stone Age. They concern the properties of the magnet in attracting iron, and of amber (*electrum*) in attracting straws when rubbed. Amber itself was an extremely valuable commodity – not so much for itself as for magical purposes – so that in ancient times it was traded all over the world from its origin on the Baltic shores.

People could not at that stage get much further with optics than in observing that light travelled in a straight line and that there were shadows. But shadows are extremely important: they are also very magical. A shadow is a picture and, in fact, I think a great deal of the development of pictures must have arisen from the study of the shadow, that is, from the silhouette.

Now, we have been looking at Table I horizontally. If you look at it vertically, you will see the periods are divided up into many rough sections according to time. I shall follow these sections as these events occurred, that is, the ancient

TABLE I

Period	Mathematics	Astronomy	Mechanics	Dynamics	Pneumatics and Heat	Magnetism and Electricity	Optics
Ancient and Classical 4000 BC 500 AD	Arithmetic Geometry	Movement of heavens Shape and size of earth	Spring bow Lever Wheel Pulley Wedge Screw	Resisted motion Sound as vibration (Pythagoras)	Bellows Pipes Pumps Archimedes Principle	Magnet and amber	Shadows Mirrors Plane and curved
Mediaeval and Arabic 1450 AD	Arabic numbers Algebra	Navigational astronomy	Horse harness *Gearing* Water and windmills Clocks Pumps	Motion of projectiles	*Gunpowder*	*Compass*	Lenses Eye spectacles
Renaissance 1600 AD	Equations	*The solar system* (Copernicus)	Parallelogram of forces (Stevinus)		Pumps for mines	*Laws of magnetism* (Gilbert)	Perspective
 1700 AD	Analytical geometry (Descartes) *Calculus* (Newton)	Elliptical orbits (Kepler) Satellites (Galileo)	Elasticity (Hooke)	Pendulum Law of fall (Galileo) *Laws of motion* Gravitation (Newton)	*Vacuum* (Torricelli) Barometer Gas laws (Boyle) Thermometer	Frictional electricity	*Telescopes* Microscopes Velocity of light Colour *Double refraction* Interferences
 1800 AD	Differential equations	Solution of longitude problem	Strength of materials (Smeaton) (Coulomb)	Generalisation of mechanics (Laplace) (Hamilton)	*Steam engine* Specific and latent heat (Black) Condenser (Watt) Heat from friction (Rumford)	Conduction (Grey) Electricity (Franklin) Condenser Laws of force (Coulomb) *Batteries and Currents* (Volta)	Achromatism
 1890 AD	Harmonics (Fourier)	The stellar system Nebulae (Herschel)	Structural calculations Fluid motion Turbines		*Mechanical equivalent* (Joule) Second law of thermodynamics (Carnot)	Electromagnetism (Ampere) (Faraday) Telegraph Dynamo *Maxwell's equations*	Polarisation *Wave Theory* Photography Electromagnetic theory

and the classical, which I shall be talking about in chapters 2 and 3 then on to the medieval and to the Renaissance – the great and most exciting part, the so-called Scientific Revolution which occurred in the fifteenth and sixteenth centuries; then going on to the seventeenth century, a more sophisticated period; and, finally, to what is really modern science, that is, Newtonian and post-Newtonian science up to 1890. I have drawn a line at 1890 because after then we move into really modern science, what we may call twentieth-century science – atomic, quantum and relativity science. I have tried in this brief survey to show you the main sequence of events that led to the discoveries in different parts of physics and brought them together.

The origin of human society

First of all it will be necessary to sketch in a little of the basic history of the times and to show the special character of the first part of the series. We will start from the origin of human society – a date, incidentally, which has never been agreed because every time an attempt is made to ascertain it new facts emerge to show that it goes further and further back. I made mention in my last lecture of some evidence of Old Stone Age counting; two basic inventions had been made. We can call them physical, though one of them was perhaps rather more chemical: they were tools and fire and both certainly go a very long way back. Peking man, or his contemporary recently found in Hungary, who lived about 300,000 to 500,000 years ago, definitely had fire. You can date him because fire leaves carbon, charcoal, and the charcoal can be dated by Carbon 14. Then even older deposits, which go back to well over a million years ago, are found in Central Africa. It is true there is no evidence that they had fire, but they did have definite flint implements.

A flint implement is an extremely interesting thing in itself because it is not only a piece of material that is used for certain purposes, but it is what might be called a social fossil. The fact is that we find the same kind of flint implements ranging all the way from Peking to South Africa and many places in Europe: and they were all made in certain ways. Now, the habits of making flints change. This enables the anthropologists and archaeologists to date a flint implement by its appearance, whether it is made into what is called a core implement by taking a stone and chipping bits off it, or whether each chip is made into an implement of its own, a flake implement. Later on they stop using the 'bashing' technique – you press the stone hard and flake bits of it off and in that way you can get the most elaborate kinds of flint work.

The point I want to make here, however, is that the technique of making flints, over very long periods, had to be taught in a school: it was as much a social thing as reading and writing. Every child had to learn how to make the flint implements and they had to learn how to make them in the right way. Now, the problem which has never quite been solved is how they ever managed to improve the techniques, because the early flint implements are, for their purpose, particularly effective. But their purpose obviously gradually changed and the material too gradually changed.

We know a great deal about flint implements for the most elementary reason that we can still find them, they are practically indestructible. But we have to assume, from what we know of other cultures, that with the working of flint implements came also the working of wood. In fact, undoubtedly, far more wood was worked than flints but all the wooden parts have disappeared. You must imagine a very elaborate social culture, very regular, and a family arrangement that went with it. One thing that is quite clear is that, although mankind gradually spread, it was spread very thinly, simply because a family in the state of nature, hunting and food gathering over wide areas, not specially favoured, needed about a ten-mile radius to live in. A small family of about ten people could only pick up a living over such a radius and, if they could not do it in one place, they had to move further on.

Another point about the Old Stone Age is that the people were entirely parasitic on nature, that is, they had to follow the animals and if the animals did not do well, the people starved. Sometimes they were extremely lucky: one specially favoured place was in North Hungary. Every year the mammoths used to spend their summer in grazing grounds in Poland but had to come back to the relatively warm plains of Hungary for the winter. There were only about three passes they could come through and people waited for them there and killed as many as they possibly could. They did it so abundantly that they made their fires out of mammoth bones. This they did by turning them so that the marrow would run out and feed the fire. They did not bother to eat the marrow, they had too much meat to eat. One of their great problems – a chemical rather than a physical problem – was how to store food. They did find many ways of doing this, but that stage, when man was dependent primarily on the animals and very little on the plants, was a stage of very slow development and only a rather limited set of techniques were evolved. But these did involve a great deal of effectively practical physics.

Weapons

One result was the evolution of weapons. The earliest weapons were just held in the hand; then a piece of leather or something similar was put at the back of the wrist to prevent the weapon cutting the hand holding it. Then a real handle was developed, a piece of horn or wood, and from this came the two basic tools – the axe where the blade was at right angles to the piece of wood, and the spear where it was in line with the handle. That development was initially designed for holding and moving and, later, for throwing. It was in relation to throwing that the first really *dynamical*, deliberately thought-out experiment came.

Let us assume that throwing itself is pre-human – we know of quite a number of apes and monkeys who throw things about more or less accurately. Yet, throwing can be improved and made perfect without any knowledge of any kind; it simply depends on practice. Even today, the principles of bowling a cricket ball and making it 'break' involves the most subtle aerodynamics which were worked out – and then only partially – only a few years ago. All the tricks of bowling a cricket ball, which depend on the slip stream around the seam, and so forth, were worked out entirely by plain practice alone. Human skills came before human knowledge. So let us say that throwing is a reasonable skill to have come naturally, but men noticed very soon that the capacity for throwing depended on the length of your arm, so that the first idea was to lengthen the arm by using a throwing stick. For this, your spear would be balanced on another stick which could then be held in the

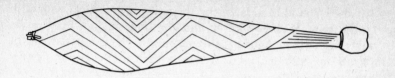

3 Spear thrower

hand. These 'spear throwers' have been found: they usually have a pin in them and this pin fits into a hole at the back end of the spear (Fig. 3).

Now, this throwing stick is the first stage for a mechanical type of propulsion and the first mechanical propulsion that we know is the bow. We do not know the history of the bow, but we can speculate a little. It looks as though the bow had two origins: one of the origins is in the use of string together with wood. String was used fairly early on: it would either be a piece of vine or liana found in plenty in some places or, in the plains, some piece of leather would be taken from an animal skin. Such thongs were certainly used – the Eskimos use them to this day. The string was then tied on to the stick. The first way it was used seems to have been with a stick in the ground – it may even have been a tree; you bent it down and tied a string with a loop to it and there you have the noose trap for animals. Old Stone Age man certainly trapped animals. Then the idea was produced of a string held tight and swinging back: you have some kind of catch and the string was held taut (Fig. 4).

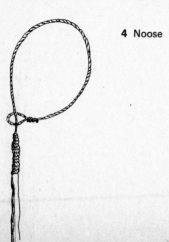

4 Noose

It was this kind of idea, of the springing of the stick, which led to the bow. Well, we know that the bow did not exist in the early Old Stone Age: it came in only in the latter part of the Old Stone Age because it was from then that the first pictures are known of the bow and of the men hunting with the bow.

Before they had the bow, men had another dynamic instrument which was really quite a complicated one and that was the bolas. There were three stones attached to the string: you held one of them, swung the others round it and then you let the lot go (Fig. 5). This was an ideal way of catching animals in the open because the string tangled itself round their legs and brought them down. But it had extreme limitations – you could only use the bolas in the open plains – if there were trees or bushes around, your bolas would just catch in the branches. We know they had bolases even though we have never found one, for the stones with grooves on them have been found.

These are some examples of the evidence of existence of mechanical contrivances, quite elaborate ones, which developed the idea of the spring, the principle of the bow and arrow, which is really a machine for storing energy and letting it go again. Then, with the bow and arrow, came something else – it may even have come first – a bow that could do something further, a bow that could make a noise. And there you have the basis of the stringed instrument and its musical counterpart. I will come later to the corresponding developments of the wind instruments and the drum: all evolved from simply using the medium and then finding out what other things it could do.

5 Bolas

The other element which I have mentioned was that of fire. Now, you may say that fire is chemical, but fire exists. Man did not invent fire, he captured it: the old legends had it that man snatched fire from heaven. The idea of taming fire, of being able to keep a little bit of fire, is very, very deeply ingrained. The fire-tamer, the fire-guarder, was the most important person in the group. If it lost its fire then the group was at a great disadvantage.* The next problem was how to make fire. Making fire has so many different forms, all of them based on a real knowledge of the transformation of mechanical energy into heat, either by rubbing and sawing, or by drilling, or by striking. And here, you see, you have the first beginnings of chemistry and also of mineralogy, because there are two fire-stones. The first they knew very well: it was out of this they made their instruments – the flint. You can manage with flint alone, but to get good results you must have a metallic substance as well. Of course, it was a long time before they made metals, but they found one, the firestone – pyrites – and it is still called that to this day. It is the iron sulphide which will yield you fire when struck by a flint, and it is a very sacred stone indeed.

Now, with fire comes the whole of chemistry and, in fact, chemistry is what fire was called in a book in the eighteenth century, *Pyritologica*, that is, the study of firestones and everything you can do with fire. One of the first things, from the scientific point of view, that can be done with fire is to make metal; but here I am running ahead because some of the things I am going to talk about first were there before metals. If you look at the assembly of Old Stone Age techniques, you find it pretty complete in relation to actual life, apart from some use of metals by the Eskimos. Everything was done that could be done with the kind of materials that were available and the kind of knowledge that could be gained from their use. People were then not very concerned with the sun and the stars; they were concerned, rather, with all kinds of mechanical devices for

*An amusing example of this, showing how late remained the difficulty of making fire, occurred when the Church and King mob in Birmingham went out to wreck the house and laboratory of Dr Priestley. They found when they got there that they had no fire and had to send out for it, and by that time Priestley escaped. This was in 1791, on the anniversary of the French Revolution.

6 Old Stone Age paintings in the Font-de-Gaume cave, Dordogne, considered by some to represent summer huts

hunting – with shaping, drilling and all kinds of things that can be done with materials, tying them together, sewing them together, putting them together in various other ways and, then, going further into a kind of architecture, so that there were various kinds of huts emerging (see Fig. 6. One of the paintings we are not quite sure about – it might be somebody's face! But they are generally considered to be pictures of huts.) We now know quite a lot about Old Stone Age huts because post-holes were left which have since been dug up. The people who made them must have had the means for getting beams and so forth and securing them together – fitting them or sewing them together. They also developed the 'water' side of this, first in the floating raft and then, by hollowing out the logs, producing a dug-out (Fig. 7). It would appear that the dotted lines on the figure are a little extra bit for using when sailing came in, the outboard or outrigger. Then, by building up the sides of the dug-out with flat pieces, they gradually developed towards the boat as we know it – which was not quickly built: it was butt ended and sewn together, and had a keel. Even today we have a keel to a boat although, strictly speaking, from the mechanical point of view, it is not at all necessary.

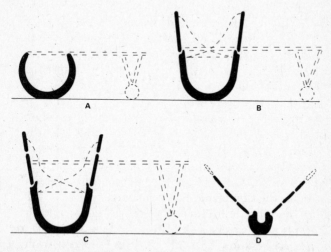

7 Sectional diagrams showing how the keel developed from the dug-out: (A) dug-out; (B) dug-out hull with one row of side-planks; (C) as (B) with two rows; (D) full development of the keel. In (A)-(C) outriggers and frames are indicated by dotted lines which have been omitted in (D)

The boat is important because the boat in use – first for fishing and then for trade – is the first man-made structure that is required to stand quite strong dynamic forces from the waves and the wind. The knowledge of how, first, to make a boat, second, to paddle it and third, to sail and steer it, is really a school, not only of the mechanics of boat building but of the arts of navigation – which leads straight up to the sky again with the principles of oceanic navigation. Its advent was certainly pre-literate, because the Polynesians used it when they sailed over very wide stretches of ocean.

The Agricultural Revolution

Now, these were, so to speak, the primitive arrangements. The whole system was rapidly transformed by a great discovery, an essentially biological discovery, that was made at a date which is being continually pushed backwards. When I first began to give these lectures around 1948, I could mention fairly safely a date of 4000 BC for the Agricultural Revolution, but this is constantly being revised. Map 1 shows a picture of the ancient world with the main places and the main conditions that gave rise to this new trend. About 7000 BC or 8000 BC the last Ice Age came to an end. The glaciers which had covered most of Northern Europe had, except in the High Alps and Norway, all melted. There was an enormous amount of open land and, of course, a much warmer climate. The whole ocean rose, as well, because of this melted ice. In another 2,000 or 3,000 years it will be about 300 feet higher than it is now, because there is still a certain amount of ice to melt from the last Ice Age – but we will not be here then so we need not worry about the consequences! The essential thing was that, with the warmth, large areas, such as the Sahara Desert or the Arabian Desert, were under water; there was a time when these were open steppe lands carrying lots of animals. Most of the animals died but some survived – right in the middle of the Hoggar there are still crocodiles, about 2,000 miles from the sea and with no rivers at all, nothing but desert all around. But once, as we know from the pictures found on walls, it was

Map 1 The beginnings of civilisation. The map shows the main places in the ancient world where the origins of agriculture and of civilisation took place. The principle Bronze and the Iron Age cities are shown

46

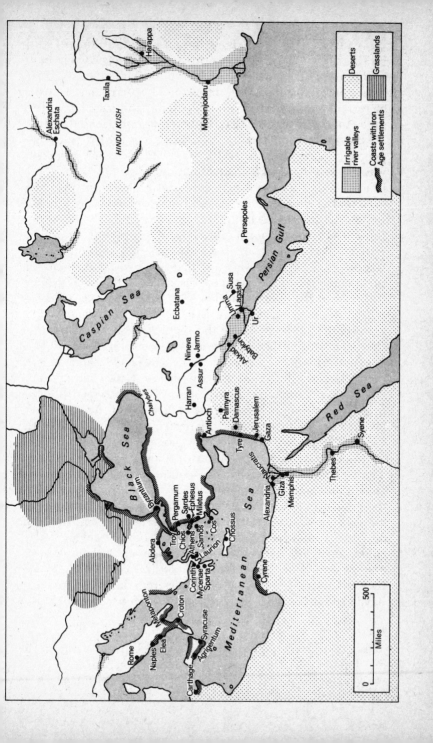

big game country with lions, giraffes, buffalo and so forth. So this drying up of the land drove people towards the banks of the rivers and it drove them also to the characteristic places where animals congregate, for example, where a small stream or torrent comes down from a mountain and loses itself in the sand. Just before it does that, however, it makes a kind of delta, a kind of cone, where the animals congregated and there people congregated too and it was there that they found the way of creating a new food supply by means of corn. It was from this that all the other agricultural developments came.

Actually, there is some dispute at the moment as to where the Agricultural Revolution actually started. Some people think it started just north of the Tigris, others that it started at Jericho. Both places are certainly very old indeed, at least 10,000 years old, and we know that there was corn there and that people ground it. We have the sickles with which they cut it, we have the grindstones and we have, up to a point, the pots in which they cooked it. In Jericho there is what is called a pre-pottery Jericho where they had clothes, they had corn – they were still in the New Stone Age – but they did not have any pottery. At another place, called Jarmo there is even earlier dating. But there was a limit to the number of people you could maintain on an area of this size. Once the technique had been developed, from these rather out-of-the-way places on the edge of the deserts, it spread into the areas along the rivers. The reason agriculture did not start along the rivers is fairly obvious; there are the three main rivers, the Nile, the Tigris and Euphrates (the Mesopotamian river) and then the Indus; and you would have your river with plenty of water in the middle of a desert – what was originally called a corridor forest – not suitable for very much except river hunting. But once the technique of agriculture had been developed, they could cope with the rivers and gradually clear them of forest and swamp. You can follow agricultural civilisation up the Nile by noting how many crocodiles remain. The crocodiles have been driven further and further south up the rivers.

The balance

This great development of agriculture brought with it the development of other aspects of practical mechanics. Fig. 8, a wool factory in Greece, illustrates two different things – here

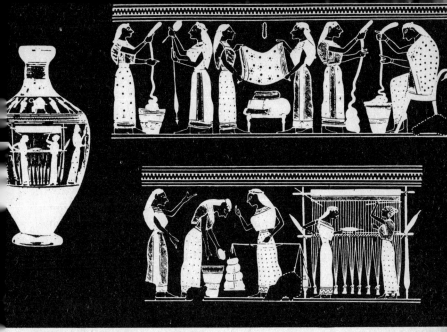

8 Spinning, weaving and weighing wool in a Greek wool factory

it is actually somewhat later, they were doing it much earlier than this – you can see them with the raw wool and then see them spinning it and then weaving; and, and this is the interesting part, at the centre of the lower part of the figure, they are weighing it. In the last chapter, I mentioned the measurement of corn – you can measure corn by volume but not wool. You can measure the length of wool as yarn after spinning if you wish, but if you really want to know how much raw wool there is and how far it will go, you have to weigh it. So you can see this primitive balance. You can also see how the idea of weight was arrived at – from a *balance* which was originally just a pole with a load-carrying pan at each end. When carrying a load on a pole, you naturally moved it to and fro to get a balance and people knew that if the loads were equal then the balance was equal because you could turn the pole right round and it would still balance. Incidentally, that balance became a symbol for everything that was correct and just – justice really is a proper balance, an equilibrium, equal weight, and we still have that notion of equilibrium very strongly implanted. Of course they did not always use an

49

equi-arm balance, that came rather later. Fig. 9 is a Roman balance of the very familiar steelyard type – a kind of butcher's scale – with the uneven arms and the adjustment of the weight by divisions. This was the common-place weighing machine of the classical world.

So far I have talked about some of the mechanical properties which imply the use of the lever. The lever finds its numerical expression in the balance, especially an unequal-armed one. The lever is as old as civilisation, it certainly must have been used in the Old Stone Age if only for poking things and lifting things. But the sophisticated use of the lever comes with the development of the balance and with two other developments connected with movement which followed, or may even have preceded it slightly. One was the development of how to move things about: the first way of moving things about, apart from carrying them, was sliding them. There would therefore be a sledge of some sort or another, and sledges were certainly used throughout the Old Stone Age. But the sledge is only suitable for rather even country and, further, it is a very wasteful thing; that is to say, sledges wear out. They will last on snow but they will not last very long on sand or rock. Something better than the sledge was required, and was provided by putting rollers under the sledge, so you get the concept of rolling. Again, that concept itself must be very old because people rolled things about, for short distances anyway, just by pushing them over and over. Logs were certainly rolled; and then logs were placed underneath as rollers – you can see pictures in museums of how it was done – which was a very tedious business. You rolled the heavy object on these rollers and you set people picking up the rollers at the back as the object came off them, carrying the rollers round to the front end and putting them under the object again. You had hundreds of people shifting enormous statues and so forth.

The wheel

The idea of the roller is well-established, but the transition to the wheel, the second development, is a more sophisticated thing, and it is really rather a difficult thing fully to understand. It is really a trick or, rather, there are two tricks involved. One trick is that you have a roller under the load and, as you want

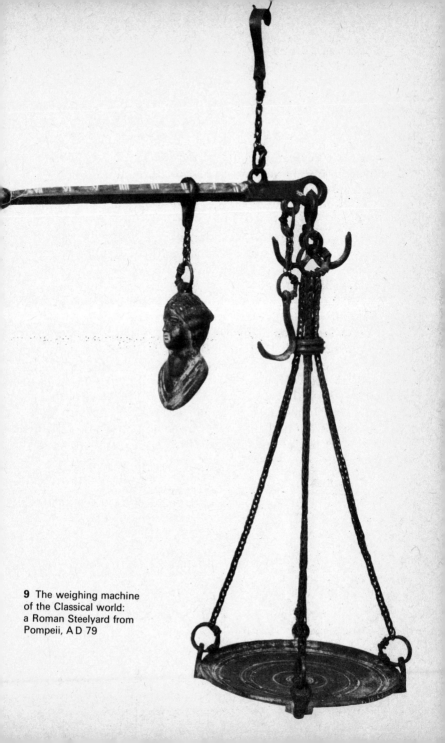

9 The weighing machine
of the Classical world:
a Roman Steelyard from
Pompeii, AD 79

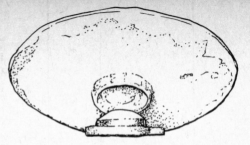

10 The horizontal wheel: turn-table and base used by some African potters today

to prevent the load running off the roller, you have to secure it in some way by putting a band round it. But that does not help you very much if the band is the same diameter as the roller, because you then have to have exactly the same amount of slip on the band as on the object moving along without a roller. But if you now put a larger wheel at the end of the roller or axle, the resistance at the axle is very much less and you can then have an economical way of moving. And this is exactly what happened.

Let us consider the idea of the wheel. The earliest and most elementary wheel is the horizontal wheel (Fig. 10); it is pivoting something and turning it round, just a stone resting on a pivot or a potter's wheel or the sockets to a door-post. The idea of putting a wheel the other way round, so that it could be used for moving things came comparatively late, though still fairly early, at about 3000 BC. In Fig. 11 you will see the first form of a wheel, a more or less solid wheel. Actually it was made out of three pieces of wood because you could not usually get one piece big enough for it. The pieces were held together by two slats and, in the case illustrated here, the tyre

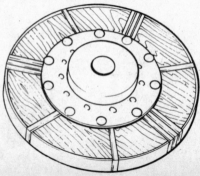

11 Nail-studded wheel from Susa, Mesopotamia, c.2500 B C; diameter 0·75 m.
12 Reconstruction of a copper tyre on a wheel from Susa

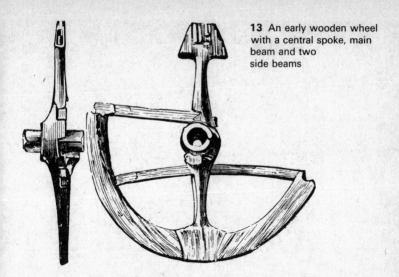

13 An early wooden wheel with a central spoke, main beam and two side beams

is in fact made out of bronze nails. Another similar one is shown in Fig. 12, here with a copper tyre arrangement. The simplest derivative of this, which was not quite so heavy, is shown in Fig. 13. It had a central spoke, a main beam and two side ones. Now, those wheels really are almost universal. In Fig. 14 there is an illustration of one being made today. I have even seen them with rubber tyres on them in Inner Mongolia. Yes, they are still being made: I never expected to see the old solid wheel in use but I saw it in Chile of all places. Of course, it was not invented in Chile, but it was all that the Spanish settlers had and they are still using it there.

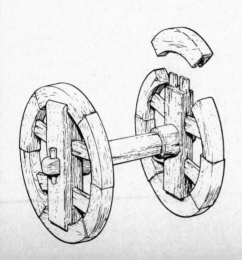

14 Wheels still being made in China that illustrate an early stage in the evolution of the wheel

15 Copper model of a chariot drawn by four asses, from Tell Agrab, Mesopotamia, third millenium B C

The use made of that kind of wheel is shown in Fig. 15. It is a little model of a third millenium four-ass chariot; the ass was used before the horse as it was easier to manage. As you see, there is a central beam, a pole, which is really the same as a plough beam. How the chariot evolved is probably as follows. To produce an implement to till the land, first you take a piece of wood with a branch or root to which you tie a stone or, later, a metal hoe. You can then either hit with the hoe or drag it along. At some stage, by fitting a yoke to the pole, this becomes a plough. You can then add wheels, remove the part not being used and turn the plough into a chariot. These things were developed essentially through use, stage by stage, without any particular sophistication.

On the other hand, Fig. 16 reveals a fundamentally different use of the wheel, although it is rather obscurely illustrated here: it is the pulley wheel with a rope. You can see how it could be used for pulling water from a well; in this illustration one of the besiegers is cutting the rope and preventing the besieged getting water, but the essential feature of the pulleys

16 Assyrians using pulleys for drawing water in a besieged town

and wheels is that they gave a mechanical force. In fact, the first elementary mechanical force was the lever; after that came the use of the wheel and then the use of a combined set of wheels or of pulleys. These things were understood practically in the ancient world and even the last mechanical force, the screw, was well known. Fig. 17 shows a rather late example from Pompeii in the first century, a press used for preparing cloth: you notice that the screws were very carefully arranged with the right- and left-handed screw to prevent any kind of swinging round of the platen.

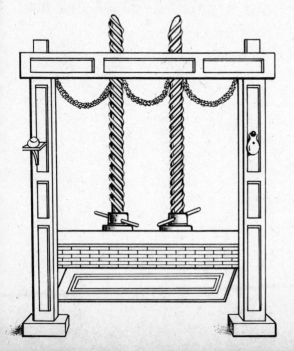

17 Cloth press, from a mural painting in Pompeii

18 (A) The Bosco Tre Case wine press at Pompeii; (B) in diagram

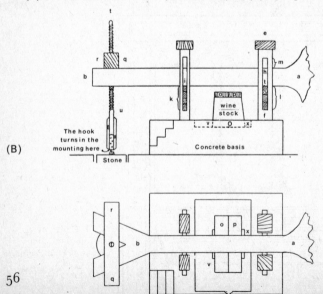

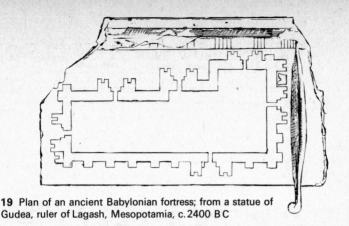

19 Plan of an ancient Babylonian fortress; from a statue of Gudea, ruler of Lagash, Mesopotamia, c. 2400 B C

Fig. 18 is a rather interesting example of a mechanical wine press which – I do not know whether it is still there – was certainly known to be at Pompeii at the beginning of this century; it is an actual classical machine which has not been altered. I have no doubt that some screws have been replaced, the big weight featured there certainly has: the main beam has been replaced but it is, in a strict sense, the same machine. It has never been redesigned, it has only been patched up. As to the mode of operation, the business part is not shown very clearly but you will note that there is an enormous weight. It is in the middle where the pile of grape skins are to be found and from which the wine is gradually pressed out. The first of the wine came, of course, by treading the full grapes down with feet: some of the juice comes out easily that way but there is a great deal of juice left and to get this you really have to press it, and the huge weight does this. The whole operation is effected through the screw and, as you can see, the screw is turned round by means of taps. It is quite a sophisticated piece of mechanical engineering and these things were much used both as wine presses and for other purposes, such as, for instance, the screw mechanism for raising water which I will talk about later.

Now we turn to an even more sophisticated side of the mechanics of the period. Fig. 19 is, in fact, an ingenious combination – a combined plan and elevation of a city and there are many cities like this. It is a fortress with gates and their gate towers, rather a lot of them. But the interesting thing is that at the top there is a set of proportional scales and it is obvious that

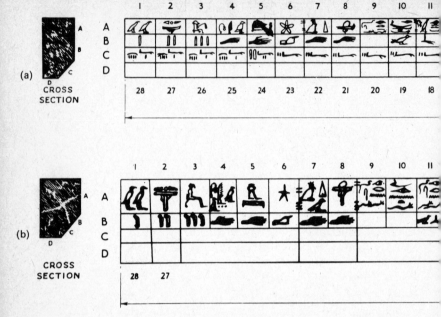

20 Egyptian measures of length. The ancient Egyptian primary unit of length was the cubit — the length of the forearm. Thus the hieroglyphic sign for the cubit was the forearm and all the sub-divisions of digits or fingers, palms, great and little spans and the foot can be seen in these royal cubits of Amenhotep I (a) and of his vizier (b) c. 1500 B C

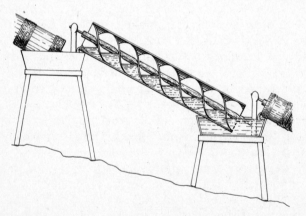

21 Archimedean screw for raising water. A whole series of such hollow screws could be geared together and used for delivering water up a considerable incline

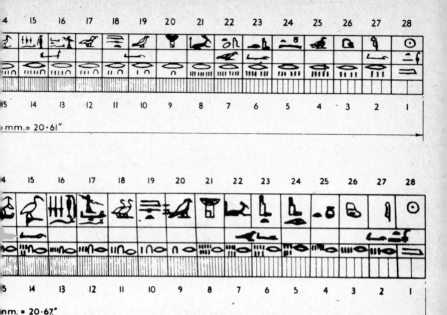

the whole drawing has been done to scale. Fig. 20 shows a plan on a rather more elaborate scale, an Egyptian variety, which shows the divisions and also explains the Egyptian fractional system giving the elementary fractions with one as a numerator, a half, a third, a quarter, a fifth, a sixth and so forth. The Egyptians worked with fractions with one as the numerator and made up other fractions by combining them and in this way they were able to work to scale. Fractions were used, too, for astronomical and financial purposes. The 'names' of the numbers were written above in hieroglyphics. These scales are, in fact, very well known: many can be seen in the museums.

Fig. 21 illustrates a rather interesting use of a mechanism to which I referred earlier, the Archimedean screw. The main use for these screws must actually have been in mines, and many such devices are found in the Roman mines in Spain. Much use was also made of all these water-lifting devices for irrigation, a way of increasing enormously the food supply in those regions where there were big rivers and plains rather above the river level. It was these conditions which gave rise to all this ingenuity and construction.

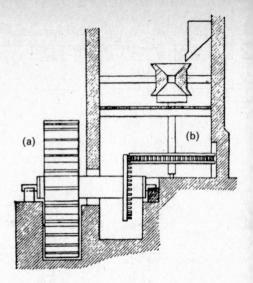

22 Roman water-mill with gears: after Vitruvius

(a) (b)

Water mills

Fig. 22, on the other hand, shows the reverse process. It is strange that we have no real drawing of water mills, but we certainly know they existed: there is plenty of literary evidence such as a very accurate description in an almost fundamental textbook of architecture by Vitruvius. Now, the interesting thing about this illustration is that it shows early gearing; you can see the horizontal-axle water wheel (a) and the vertical-axle mill above (b). Actually, it started in the reverse way with a vertical axle, with an ass or camel walking round it. The wheel, instead of being a wheel moved by the water, became a wheel that lifted water – a set of buckets lifting water, and they could work a mill simply by putting that process into reverse.

The working of mills gave rise to the first real use of energy other than human. If you look at some of the old Egyptian relics, you will see how milling was the most exhausting business. In the palaces of the Pharaohs you would see any number of women all kneeling and grinding their grain on a hand grinder – there are models of thirty or forty women at a time. Now, the principle of the mill should have stopped all that but, in practice, it did not. One very practical reason for this was that in most of the countries where civilisation had developed, water supplies were somewhat erratic – they would dry up altogether in the summer – and relatively little practical use could be made of them. Curiously enough, however, the mills were used even where these conditions seemed to apply.

The earliest description we have of a water-mill – and in great detail – is from some Irish laws laid down in the third or fourth century AD. The mill referred to there was a much simpler device, it was a vertical mill, a turbine. The water fell on the turbine blades at the bottom of the well which were set at an angle and the mill was mounted directly on the axle; there was no gearing. Mills of this description could be set up all over the place, and in fact they were. In the Domesday Book, some 5,900 mills are recorded for England alone in 1086. The mill gave an enormous advantage to places where there was running water, and you will see how important it was much later in the Scientific Revolution and in the Industrial Revolution. It is not an accident that William Blake talks about 'dark, Satanic mills' – there had to be mills in just those places because they were the only places where you could find power.

Pneumatics

I will now go on to one of the more sophisticated aspects of mechanics, pneumatics: all the properties of gases or winds. Wind, in a sense, is the oldest form of power, although not necessarily recognised as such because the first application of power was not an elaborate, mechanical one at all – it was simply the sail boat. The wind in the sail drives the boat along and although people had not developed the art of sailing then into what it is now, they appreciated the fact that in general they had to sail with the wind. So they depended on a favourable wind and their main problem was how to raise the wind in one way or another. Sometimes they used somewhat drastic methods. When the Greek fleet was ready to start for Troy during the month of August, there was no wind – and still no wind: something really had to be done. So Commander-in-Chief Agamemnon's daughter, Iphigenia, was sacrificed in order, as he thought, to ensure a good wind. He got his wind but subsequently he was murdered. The story goes on of Odysseus cornering the wind god and getting all the good winds stored for him in large, what might be called, sacks, although they were probably really skins. He then had skinfuls of wind which he kept in the bottom of his ship only letting out the winds he wanted. But, unfortunately, the crew, being very suspicious – and they had every right to be – thought he must have some treasures stored in the skins. They opened them up and all the winds came out; immediately there was a terrible storm and Odysseus's ships were all destroyed.

Of course, you would not expect much theory in matters of that sort. But there was another wind, other, that is, than the winds of Nature, that could be relied on: it was your breath – you could blow. Breath was a terribly important thing, not only for mechanical purposes but because breath, after all, was equated with life. You had the breath of life and as long as you were breathing you were alive. It was considered, therefore, that this could work in reverse: you could breathe life *into* something. You still can; it is the best method even now of life-saving after drowning; you give the drowned person back his breath by breathing into his mouth.

This was associated with another use of wind, that is, blowing *at* various things such as blowing at the fire with bellows. Now, the bellows is the basic invention that made metallurgy possible. If you try to make metals on a fire which is just an open fire burning by itself, it would be practically impossible, but once you introduce the bellows you can get a much hotter fire, and you can melt metals. A skin bellows would have been the first kind used, just a skin which you could press and lift with a string and go on pressing. That was the origin of the great metal revolution.

There is still another way of blowing – with, as it were, two more uses. One, of course, is to blow an instrument, to blow a pipe, and we know they blew pipes even in the Old Stone Age because the earliest pipes were made very widely, out of bone, hollow bone. There is a picture that I saw in one of the caves in Les Ezies in France, of a woman blowing a horn – a very primitive thing to do. I came out of the cave, a rather obscure cave, I might say, and there was a farmhouse nearby where I saw the farmer's wife blowing a horn to call her husband home to his meal. Things had not changed at all!

Now, the blowing (wind) instrument gave rise to the development of music and later still to the question of the theory of sound. The theory of music in the West was based mainly on stringed instruments: in China it was based very largely on stone and wind instruments.

The final use of pneumatics, however, turns out to be the most important of all, that is, blowing as a form of propulsion – the blow gun, which is still in use, is essentially a forest weapon just as the bolas is essentially a plains weapon; in the blow gun you have a gun barrel and a dart, usually a poisoned dart. The Greek word for arrow, incidentally, is 'toxon'. The Greeks would not waste an arrow; if they could put poison on it, they did, in war, of course, and in everything else, as the story of Philoctetes shows. From the blow gun, then, came the idea of moving an object along inside a cylinder. This concept of the pneumatics of a cylinder gave rise to an air pump, which may be a pressure pump pushing air out or a suction pump pulling air in.

Fig. 23 shows one of the rather sophisticated developments – I have seen one of them which was dug up in Hungary – it is a Greek water organ. The water part of it is not so terribly important here. It is first of all a real, proper air pump, no different from the one you see every day providing compressed

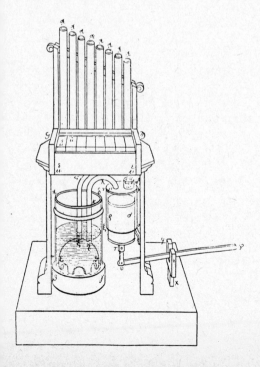

23 Greek water organ

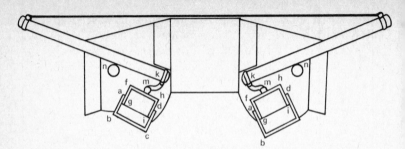

24 A pneumatic catapult operated by compressed air: attributed to Ctesibus, c.1st century BC

air for the drills in the road. It is maintained under a constant head of water in one pipe and finally the air comes up through another pipe to operate the organ. You will notice the keys – those keys were known probably in the seventh or eighth centuries BC – it is a keyed instrument and, consequently, when they came to make the piano, they simply based it on the organ keys because everyone knew about playing organs. But that is one sophistication: another one, as you will see later on, is a military use of the same conception.

Fig. 24 shows an idea for a pneumatic catapult. You can see the string, the rods and the cylinders. The cylinders operate the rods – they have to be blown up with some kind of pump – and then let them go. I am not at all sure that this device was ever actually used but there are certainly pictures and descriptions of it. We do know, however, that pumps were used – the ordinary suction pump and the force pump for water and even the double-acting force pump were all fully known. The problem really, when you come to think of it – and we will discuss this later – is how, with all this knowledge and sophistication, the Greeks did not go much further than they did. But you will see from the next chapter why this was so.

3

The Origins of
Classical Science

In this chapter I propose to try to trace the origins, development and interpretation of the practical ideas that I talked about in the last.

First, I shall say something of what might be called the world set-up at the period when science was born. There were at that time – at about 2000 or 3000 BC, it is very difficult to fix this more precisely – three more or less mutually isolated centres of civilisation, places where the devices that I described earlier were being used. These were, in the first place, *the Mediterranean area and further East*, the so-called Near East going down to the Persian Gulf and including primarily Egypt, Syria, Mesopotamia with certain outlying territories; *India*, particularly the Indus valley, which is now largely Pakistan; and *China*, which was much further away from the others. Map 1, p. 47, shows two of the main centres, first the Near East, the so-called fertile crescent which stretches from Egypt to the Persian Gulf. Then, really quite far away from it, separated by more deserts and a large area of sea, is the second very similar civilisation in India. The other places marked as irrigated all have the same characteristics, they all present the second stage of agriculture, that is, irrigation supplied by rivers in the mountains and going through desert lowlands, such as, for instance, up in Kwarism and down in Armenia. This kind of civilisation is characterised by the early growth of fairly large cities all of which involved quite elaborate administration, leading to the use of mathematics, considerable building, which I mentioned earlier, and with that the materials for the origins of science.

The conditions in China were essentially the same, that is, civilisation was mostly located around the basin of the Yellow River in the north with irrigation, and only gradually spread into the forests and wetter country of the south, along the Yangtze and in Kwantung. At this early period, which is called the Shang period in China, there were typical city civilisations; in fact, civilisation is the city, that is what it means – *civis*, the city.

The calendar

Now, you may ask, where does science come into this? Essentially, it comes into the carrying out of the necessary routine functions of city existence and the most scientific parts of that routine are the ceremonies concerned with production, particularly agricultural production, which involves the calendar. In fact, in early days the history of the calendar is very much the history of science. The calendar involved the knowledge and understanding of heavenly motions and this was considered extremely important. Obviously, some of it was important. The Egyptians were particularly concerned with it because the Nile flood was a regular annual event. Other rivers were very unreliable in their flow; you could never be sure what they were going to do and occasionally they overdid things altogether and produced floods. You may notice that Map 1 is not quite the same as the modern map of that area. The sea has now receded some 150 miles from the city of Ur, which was then a seaside town. In Mesopotamia there would be terrific floods and everything around would be inundated. The Bible account of Noah's Flood probably gives a fairly accurate description of one of these floods. They left plenty of mud and gravel behind which is still there.

In Egypt, on the other hand, there was a regular flow from the equatorial rain forest and a periodic flood which came as a result of the monsoon in Abyssinia. It was essential to know when the flood was coming and very early 'Nilometers' were constructed – one is still there – which marked the height of the river. The people who made the calculations had to give warning when the floods would be coming and the whole of agricultural life had to be organised round this event.

That was one object of the study of the calendar. Another object in the early days of agriculture was to determine the ordinary planting and harvesting dates and, what was considered much more important, to be able to be sure that the weather would be right. This was not just a matter of finding out, but of influencing the weather! People believed that by taking the proper steps and by saying the proper prayers at the proper time, they would get rain and could avoid or anticipate floods, and that it was all tied up with what was originally a combined arrangement of priesthood and kings. The latter were, so to speak, the life of the whole community and the proper ceremonial action of the king settled the fate of the kingdom.

In China, it was all much better documented; we know, for instance, that the poor Emperor of China had to get up at four o'clock every morning because, if he did not, it was not at all certain that the sun would rise. He was also very concerned with events like eclipses. Two early Royal Astronomers, Hi and Ho, failed to predict the date of an eclipse and were accordingly executed. It is very difficult to predict eclipses and they were just unlucky. However, the point is that proper observation and proper understanding of the skies were considered to be of enormously greater importance than they have been ever since. So you can see that a great deal of physics was built round the calendar, which arose out of the observation of the skies, as I pointed out in an earlier chapter.

Another aspect of physics was the study of what might be called the history of the universe. Now, curiously enough, the title for this in Greek is 'physis' – the coming of things into existence, and, in the same way, translated into Latin, it becomes 'nature', the birth of things. All early thought, both in the empires of Egypt and Babylon, and in the new areas I will be discussing next, in Greece, was concerned with that *nature* interpreted in various ways. Of course, nature had to be interpreted in the first place in terms of gods, usually in terms of the local gods, and the various legends built around the origin of the world were the basis of religious ceremony. They might have gone on for a long time being so because, of course, they were the things that people were most attached to and any interference with the beliefs of the origin of celestial things was reacted to very badly. Even as late as 430 BC, the physicist Anaxagoras was expelled from the city of Athens and very nearly executed for suggesting that the sun might be as big as the Peleponnese in Southern Greece. As everyone knew, the sun was a god and moved in a chariot and this suggestion was putting the sun at too far a distance altogether. But gradually, this attitude changed and this showed itself particularly in the new aspects which appeared further west. Now, the empires of the East were extraordinarily well established and very stable; they had domestic troubles, of course, and occasional revolutions, but by and large the way of life was extremely settled. So it was in China, where it remained that way practically until 1911. The fall of an empire was an unprecedented thing; dynasties changed, even foreign conquerors arrived, but the idea of the empire and the idea of the proper way of behaving was permanent.

About 1600 BC, for reasons which we need not go into here, many disturbances occurred. New, strange people came down with horses and chariots, people from the grasslands (Map 1, p. 47), which probably were drying up. Down they came and they had terrific damaging effects on the old civilisations. They did not destroy them all at once. In fact, the Babylonian civilisation was only fully destroyed at the time of the Tartars in the thirteenth century, and Egypt was never completely destroyed because the Tartars overreached themselves; they were actually beaten off and never got into Egypt. But the effect was to bring these tribes down on to the coast where they met the people who were there already – a variety of people who were only very partially civilised. As you can see from the position of the Iron-age settlements marked on Map 1 they obviously had a different way of living because, instead of concentrating on river valleys, they spread over coasts and islands.

The Greeks

The Greeks knew about civilisation very well as they were really part of it themselves. In the early Greek poems you are still in ancient times. Then came the Bronze Age when the existing way of life began to break up; after that came the very hard and horrible Iron Age, the period when the first poems and histories were actually written down. In the Iron Age, the chief products were iron weapons and iron agricultural implements, which enabled things to be done that could not easily have been done earlier – the cultivation of dry, stony country not requiring irrigation. This land did not produce as much, but to compensate for that you had the things that could be made with iron. In those days all the Eastern Mediterranean area was very different from what it is now, it was extremely heavily wooded. There are few trees there now, but this was one of the most important, productive areas in the world and it still retains its name but not the trees – the Cedars of Lebanon. The Cedars of Lebanon covered the whole area right up to Damascus and was an invaluable source of timber for ships.

Greece was covered with timber, also used for ships – and the Greeks certainly had a wealth of ships. With these ships they travelled all over the Mediterranean area, never going far inland, building cities along the coast, all along South Russia, along Italy, Sicily, North Africa, anywhere where there was at least some water and there they settled. Now, for us, I think the most important aspect is – how did science grow in these circumstances? It grew from a set of people who travelled around as traders. They had no particular knowledge to start with; they had to pick up knowledge as they picked up wood, here, there and everywhere. They picked up some Egyptian knowledge, some Babylonian knowledge, they even picked up a certain amount of Indian knowledge; I do not think they got as far over as China. True, their knowledge was much less perfect, much less established than that of the Egyptians and Babylonians, a fact which was rather important in the field in which they were most interested, namely astronomy. This was because the essence of astronomy is being able to take observations over a long period of time and for this it is necessary to have a more-or-less stable situation for observatories.

Eclipses

Earlier, I mentioned eclipse prediction in China and this can be worked backwards. The prediction of eclipses was made by the so-called earliest pre-philosopher, Thales of Miletus, around 560 BC, and he could not have made these predictions without tables of earlier eclipses. He predicted an eclipse that was historically rather important because it stopped the battle between the Medes and the Lydians and ultimately led to the destruction of the Lydian kingdom because, when he went back there the Lydian king demobilised his forces but the Median king did not and so took over Lydia. The point I want to make here, however, is that Thales could only have made that prediction on the basis of very long tables of eclipses going back for at least 200 years, tables which he could never have possessed himself. He must have got them from Babylonia. In fact, the observations of Babylonian astronomy must have continued right up to the time of Mohammed and later, because one particular city – the moon city of Harran – was exempted by the Mohammedan forces who had orders to attack all heathens but to spare the peoples of the Book. The peoples of the Book were, of course, the Jews and the Christians who had sacred books. But the people of Harran were able to show some kind of book of astronomical observations and they were included in that exemption. They continued to make their observations while worshipping the stars, moon and sun. Thus the actual material observations as well as calculations were made at Harran.

Calculation

All the numerical methods of calculation were carried out to an extremely high degree of accuracy, a fact which has been appreciated only recently very largely through the work of Neugebauer* who has retranslated the clay tablet manuscripts. They show that the calculations were carried out with this extraordinary accuracy simply by counting occurrences of the planets, of the moon and so forth, using a notation which was essentially a decimal notation, a place notation, but using the number of sixty instead of ten. We use this system still, we have minutes and seconds, but for ordinary purposes we do not have thirds, a third is a sixtieth of a second. Instead of having microseconds and so forth, we could use thirds, fourths, fifths, etc.

Now here Babylonian mathematics was a very solid foundation for calculus and an equally sound foundation for geometry which existed in the work of the Egyptians. How were the Greeks different? It was not that they were any better at calculation than the Babylonians, they were far worse. They had no proper number system – they used letters for numbers – and it would practically have killed them to do even what an

*O. Neugebauer, *The Exact Sciences in Antiquity*, 2nd ed., Providence, R.I., 1957.

ordinary child will do now in a multiplication or division sum. Archimedes, who was a very clever man, was so proud of himself when he found out how to notate large numbers that he wrote a book about it. In fact, the Babylonians had been notating large numbers at least 2,000 years earlier. The Greeks were bad at calculation and their practical geometry was certainly no better than that of the Egyptians. They had no positive advantages, but they had a very strong negative one – they had no traditions, no respect for the gods of the various places, because the Greeks travelled around. If you lived all your life in the city of Heliopolis, the city of the sun, you naturally had a very strong fixation on the local sun god, Ra; but if you only visited it occasionally and went on to other places, you built up a picture of the universe in which the gods were left out. If you had no gods and no proper tradition, you had to do some thinking for yourself and, essentially, that was what the Greeks did.

If we follow the trace of Greek physics (what the Greeks called physics, we would not call all of it physics now) you would see it as a gradual breaking away from old traditions and the building of other traditions out of them. The chief places where this happened were along the west coast of Asia Minor, around Ephesus and Miletus, in the region called Ionia (Map 1). From the city of Miletus came the first man of science, Thales. The culture of the area was obviously mixed, including Greek culture itself and the Babylonian and Hittite culture that came from the East, the cradle of Greek science. In the fifth century BC the Persians conquered Greece and many of the native peoples went away and founded other cities, mostly in South Italy and Sicily because the region in Greece itself was fairly congested. So the first home of science was in Ionia and the second home was in what they called Greater Greece – the Kingdom of Naples and Sicily as it was called until the founding of the Italian state, which has always had a kind of cultural unity owing to the presence of Greeks there.

Greek science

With that introduction, I should like to say something about what Greek science in fact was and how it grew. It grew remarkably quickly; all the main ideas had more or less formed themselves in a matter of 150 years. But there seemed to be a very definite sequence of ideas which spread over the whole area and this seems clearly to indicate that the Greek cities were all in touch with each other – in fact, their wise men quoted each other. The culture was what the Greeks called Panhellenic – they all spoke the same language, they all traded with each other, they were all able to compare notes, to argue with each other and disagree with each other. At the beginning it was what might be called the refinement of the world picture. Initially, the world picture was, of course, very crude indeed: it was simply this – there is the earth, up above there is the sky, and you could interpret the sky in any way you wanted according to your own techniques. If you were a plainsman, you could say that the earth was a tent; if you lived among the rocks, you could say the earth was a cave or a building with a stone roof to it; if you had progressed a bit further than that, you could say the sky was a cup over the earth; the earth was underneath, up above was the sky and between them there was the habitable world with everything in it.

The difficulty, however, already apparent to the Egyptians, was how to explain the sun. You could consider that the sun was born every morning and died every night and every morning came a new sun; that was one way of looking at it. But most people, once they began travelling, did not think this a very good idea. They thought, instead, that the sun came up in the morning over the eastern mountains and went down in the evening over the western mountains and, in between, the sun had to go on a journey under the earth in order to come up on the other side. This was the Egyptian view and you can see it painted very nicely on the inside of the sarcophagi with mummies in them. For the Egyptians, the idea was that if the sun could do this then the Pharaohs could do it too. They believed that the Pharaoh lived and died and re-emerged as another Pharaoh and, later, with the democratisation of religion, that anyone could do it, if he could afford to get himself properly mummified and furnished with the proper books and maps of the underworld so that he could find his way in and out of it again.

They were beautiful maps: each map showed all the dangerous places, had all the remarks and answers you would have to give to the customs officials and others who would try to hinder you. Naturally, there would be another little difficulty somewhere in the process, a judge who would examine what you had done and weigh you in a balance. With a simple word you could get round that difficulty too.

However, this idea of the sun and, of course, the moon, too, circulating in this way, was the first to break in the rigid old world picture. The astronomers went a little further. They said the sun and the moon were there as timepieces, they were there to determine the hours, the days and the months. You can find the precise definition of this in the first chapter of *Genesis*: 14. 'And God said, Let there be lights in the firmament of the heaven to divide the day from the night; and let them be for signs, for seasons, and for days and years.' But the Greeks did not have this refinement. They wanted to find out what the sky was. There was a very much older legend of how this happened, how it was that there was a sky above and the earth beneath. Well, the old legend, with the much cruder idea, was that there was once a universal monster, an enormous kind of whale or crocodile, and that the hero, the sun God Marduk, had a fight with the monster and tore it open: and the upper jaw of the monster was the sky and the lower jaw the earth and the sea. This you find in a slightly modified form in the second account of the creation in the second chapter of *Genesis*. The first account was obviously written by some scientist and the second account is technomorphic – there is nothing in it about seven days or anything like that – it is about making a man from clay.

Now, this is where the Greeks started – with the former idea. Thales's idea was concerned with just plain water, that water was what everything came from and it divided itself into earth beneath and air above. Water was the first element, the basis of every living thing. This did not explain anything very much, but it did at least give some kind of natural picture and, incidentally, the monster had disappeared, the sun god had disappeared: this was just nature, the way things were. Thales started this idea and it was carried on by others in his home town who were his pupils or people who argued with him. Some considered, for instance, that it was not quite like that. Anaximander said that it was mist and not water up above and that there was a kind of permanent mist over the earth.

Occasionally, there were holes in this mist and outside it there was a region of fire and light. The fire and light shone through the holes in the mist and that explained the sun, moon and stars. As you see, this was a much more refined hypothesis. Some even went further. They said that the earth was not just something covering everywhere, they knew that around the earth there was water; so to them the earth was a kind of saucer, or perhaps it would be better to say a wooden platter which just floated on the water. They assumed that it was round, simply because, I believe, they could not think of any other shape for it, and they even gave measurements for it – how thick it was, how wide, and so forth – which corresponded with travellers' tales. This was the old earth, the old world, around which flowed what they called the river of ocean. This was called a river because people who had been to remote parts had noticed that it did not stay still as a proper sea did but that it rose up and subsided at different times. The proper sea, as everyone knew, was the Mediterranean Sea and that was a kind of sea on earth. But outside there was the earth on the sea, this surrounding ocean.

The philosophers

Well, these were still primitive pictures continually being rationalised. Yet with them went a certain amount of measurement, particularly of the kind that could be made easily with the gnomon, or sundial pointer, noting days and years and following up the Babylonian calendar. The first break with this – the first complete break – was made by the Sicilian or Italian school. There, in one of the Italian cities, lived a Greek philosopher, Pythagoras, who developed the idea that mathematics was concerned with nature. Previously there had been very little thought given to a connection between mathematics and nature. There were certain magical arrangements and the people who studied these matters were called philosophers. What do we mean by philosophers? There were philosophers not only in Greece but also in India and China. Now, a philosopher had a certain special part in society, a philosopher was what would now be called a scientific adviser. He had to make his living by staying in a court or in a city and advising the ruler or the government as to what they should do in various circumstances according to the omens. Very often, particularly with the Chinese philosophers, it was a matter of manners. There was a crucial instance when one of the rebellions was successful and the Han dynasty was set up after a great many battles and much fighting. The Han Emperor said: 'I won this Empire in the saddle and in the saddle I am going to hold it', and he threw out all the philosophers. But one of the philosophers managed to remain at court and said: 'I will wait'. So the Han Emperor called a sort of grand victory feast with all his generals, who began quarrelling as to who should occupy the highest places at the feast and it came to blows and the whole event nearly broke up. Then the philosopher said very gently: 'If you had listened to me, you would have established the proper rules of conduct and of order, so that the position of every general and of every person is fixed, and then there would be none of this fighting.' So, in the end everything became perfectly ordered again.

It was for that idea of order and good behaviour that the philosopher was able to draw his money. Besides this, he could, of course, predict the future, he could cure diseases, and do a number of things that ordinary people could not do. In the more sophisticated cities, however, the philospher had to advise on the city regulations. For instance, Solon, one of the greatest of the philosophers, drew up a complete new constitution for Athens, deciding which were to be the wards, what voting there should be in each ward, how to arrange the different assemblies, how to order the state, and so forth. Now, ordering the state and ordering the world were very much the same kind of thing and it was in such a situation that a person like Pythagoras found himself. Of course, if the particular state or city fell, the philosopher either had to die with the city or take a new master somewhere else, and this happened to Pythagoras himself in the South Italian town of Croton. There he established the first school of philosophy.

Pythagoras

Until now I have been talking about individuals. There had been no Thalesians or followers of any of the other people I have mentioned. But there were Pythagoreans. Pythagoras set up a school which had very strict rules, some of them very mathematical and some of them of a kind which do not make sense to us; for example, it was absolutely forbidden to eat beans. Why beans? Nobody seems to know – perhaps because they made wind – however, no Pythagorean was allowed to eat beans. In this school, Pythagoras taught a number of things: one of them, of course, you all know, although it is the one theorem that he did not, in fact, produce! The theorem of Pythagoras was known many thousands of years before Pythagoras. In fact, there were complete lists in ancient Babylonian tablets of all so-called Pythagorean right-angled triangles giving the lengths of their sides so that they could be worked out, even when they were not whole numbers. There was a complete list of the whole number sides and tables of the square and cube roots – the Babylonians had quite a good arrangement of mathematics.

Nor did Pythagoras, incidentally, discover the proof of this theorem, that was done much later, but he *announced* the theorem and he considered it as a great mystery. Here was a fact about nature which was quite independent of man or anything else – that the square on the hypotenuse of a right-angled triangle was equal to the sum of the squares on the other two sides. And this was the beginning, or the first record we have, of the particular feature of Greek science that was different from all the other sciences that had gone before. Although, obviously, we do not have the best Egyptian and Babylonian books, but only the commonest, such as school books, and therefore we cannot be absolutely sure that they never had a discussion or proof of anything, we can find no vestige of a discussion or proof in any of the documents we do have. They simply contain rules: to find the volume of half a pyramid, you multiply this by that and take the square root and there you are – but it does not say why.

The Greeks, however, developed this idea of proof – how, nobody knows. My own theory is that they developed it because they were such acute lawyers and the terminology, the way a proposition is put and proved, is very much like a case at law and it ends with a verdict – a verdict that it is so or that it is not so. You have certain agreed statements from which to begin, you take them as axioms and you go on to build other premises from them.

But Pythagoras made another and much more important

contribution. He introduced physics in the sense that we know it now. He studied harmony and is alleged to be the first person who was able to show the relation between the sound and the length of the string. Now, stringed instruments existed long before anyone knew anything about harmony, and they were obviously tuned and they obviously had scales. But these scales, which are found in instruments all over the world, were found entirely intuitively by what might be called natural harmony: people liked the sounds like that. Pythagoras reduced the scale to numbers. He had the idea of the octave and of the other harmonic ratios – thirds, fifths and so forth – and he had the idea that this harmony was enormously important. It led, of course, to what we call the harmonic progression and, therefore, mathematics became mixed up with physics for the first time. Pythagoreanism is mathematical physics.

Another thing that Pythagoras was concerned with was the properties of geometrical figures. I have already mentioned right-angled triangles but this also applies to polygons and to the various regular solid bodies in which the Greeks became interested. As you will see later when we discuss Plato, a picture was built up of the five solid bodies as sacred things. They were sacred, perhaps, to start with, but they were also useful. The people at that time were very much given to what is called divination – finding out what was the will of the gods, finding out such things as whether it would be an unlucky day for planting fruit trees or a lucky day for going on a journey. Originally, it was another case of democratisation. There were so few scientists around that only a Pharaoh or a king could get advice of that sort, but as more schools were set up there were more people capable of doing these things and divination was more and more practised, and in various ways which are still used. It was practised by looking at things such as odd-shaped spots and by looking at animals and birds. This is an aspect which might almost be called biology. Animals were cut up and if you saw anything queer inside them you knew something terrible was going to happen. That is what is called augury. Again, if you had a good look at the birds, you could see which way they were going and you knew from that what was going to happen. Incidentally, this is all embodied in a language, not in our language but in the French. You know what 'happiness' is in French, it is *bonheur, bonum augurium*, a good

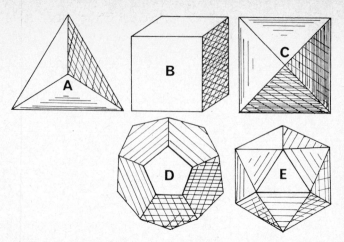

25 The five regular geometrical solids: (A) the tetrahedron, four faces; (B) the cube, six faces; (C) the octahedron, eight faces; (D) the dodecahedron, twelve faces; (E) the icosahedron, twenty faces. In all these solids the faces are equal in area and shape

augury; and misfortune, *malheur*, is *malum augurium*, a bad augury.

The way to find out these things, apart from these more or less biological ways of doing it, was by means of more geometrical objects thrown as dice. The four objects that were first used were natural objects, usually little crystals, the commonest crystal being a cubic crystal occurring in pyrites, which was, in any case, being used for striking a light. People picked a nice cube of pyrites, threw it around, and noticed which way up it came. At first this was done for religious purposes only, but afterwards, such is the nature of man, that he put bets on it and played games with it – hence the dice. Now, of course, there were various ways of doing this and you did not really need dice. You could mark spots on a piece of board or something like it, and you could play them in different ways as in cards or dominoes – dominoes is a kind of dice game. And you could have other shapes; one of them, in pyrites particularly, is the pentagonal dodecahedron, which has twelve pentagonal faces (Fig. 25). This was a perfectly good, and obviously mysterious way of making a die with twelve markings on it. There is another solid, the icosohedron, which has twenty faces.

These dice forms were, very naturally, given very considerable magical properties. There were five regular solids, no one could find any more: there was the *tetrahedron*, a single pyramid; there was the *cube*; there was the *octahedron*, two square pyramids joined together; and there were the two forms mentioned previously, making five. A great deal of attention was given to these five shapes and that led to a way of thinking which flourished very much in China and almost as much in Greece, but which was actually delusive. It is what might be called identification by classification – we will see examples of this in a moment.

The elements

We will now come back to the picture of the world. It is becoming gradually refined: it has earth underneath, on top of the earth there is water and on top of the water something was discovered – the Chinese never actually gave a name to it, nor did the Greeks for a long time – and that is air. Air, was very important but rather difficult to get a hold on. What they knew very well was wind. Now, wind you can feel and you can produce it, you can blow. But the notion of a material substance which came out when you blew or which came in when you sucked was much more difficult to form. As I have said, this conception was never formed in China but in Greece. Also, from the start, you had what the philosopher Empedocles of Sicily considered to be the most important element, that is, fire. Fire has been particularly sacred from the oldest times. In fact, it has remained sacred to a number of people in Persia and to the Parsees. Fire had one property which made it different from all the other elements. The others went down or stayed in their places, but fire not only went up, it was up there right on top – there was the fire of the sun and there was the fire of the stars. Fire was obviously the most superior, the most noble of all the elements; it was also enormously powerful. When the metals of the Bronze Age were developed, the use of fire for these purposes and some particular tricks that could be done with fire became very important. One of the characteristics of the Bronze Age was the change in the method of disposing of peoples' bodies. Before that time they were always buried, sometimes in the earth, sometimes they were stuck in trees, but in any case the bodies were just put away. But when the metal workers got going, they found that you might have an ore, a very miserable looking thing, dirty and friable, which, if heated up in the fire, suddenly went all clear and a beautiful, bright drop of shining metal would appear – you obtained your metal from the drops. This was considered to be a natural process and it gradually came to be the way in which first heroes and then ordinary people were disposed of. Their bodies were just burnt and their ashes preserved. The ashes obviously contained the noblest part of them which would resist fire.

Now, these pictures of the elements led to what might be called a conventional picture of the world arranged in four layers: earth, water, air and fire. These are the four elements. Elements, by the way, are simply letters; they are also referred to as letters, *stoicheion*, and came into history about the same time as writing, that is, about the seventh or eighth century BC. You could spell out the world simply in terms of those elements. The elements we know now are literally derived from the old ones, although, in fact, none of the ancient elements is what we would call an element now. Earth is a frightful mixture, water is a compound, air is again a mixture, and fire may be anything – I mean it is a state not a substance. But the Greeks did not concern themselves with that: they were concerned with the fact that they knew where they were with the elements. They could go further, they could even say that they could change the elements. You could take a piece of wood, you could put it into the fire, and first of all it would sizzle a bit and water would come out of it and then steam and then flame, and in the end you were left with earth, that is, ashes. So people and ordinary things must be composed of all the elements mixed up and could be decomposed mostly by the element of fire. It also explained the most ordinary behaviour of the elements, their motion. Earth, for them, was anything solid, and anything that was properly composed of earth sank in water; if it did not sink in water it was because it had something else in it – as indeed it had – a piece of wood floats on water because it has air in it. Earth will sink in water, air will rise in water. Water will fall through air, and fire will go up because the instinct of fire is to try to get where it belongs, to the divine fire in the sky; and this is the analogy.

Now we will come back to the solids of Pythagoras and, later, of Plato. You had originally four regular solids – the icosahedron was discovered last. As there were four solids it was very natural to make these solids stand for earth, air, water and fire. These were symbolic understandings – which did not mean that they actually were that, but people were apt to identify them in that way. Then, when they found the fifth solid, they naturally had to find a fifth element to fit the five regular solids because, if the elements fitted the regular solids and there were five of them, then there must be another element. So they discovered another element – which has had a rather fluctuating life and I am not quite sure where it is now – that is, ether. Ether is supposed to be something which is superior even to fire, but it has all the admired qualities of being totally undetectable by any known means. You could not catch it, you could not define it, you could not do anything with it. So we now have this elemental picture, this picture of natural motion, and with it is combined the picture of the universe itself.

Earth was for a long time, very naturally, the centre of all things. But visible things consist of the sun, the moon, and the visible planets – Mercury, Venus, Jupiter, Mars and Saturn – five planets which could correspond to appropriate elements. Going further, there were metals; there were seven metals corresponding to the five planets plus the sun and the moon. The sun being superior was obviously given the appropriate metal: the sun was gold; the moon was silver; Jupiter for some reason or another was tin; Mars, because as everybody knew it looked red, was obviously iron; Venus, again for some peculiar reason, was copper, perhaps because the goddess was a Cypriot – she came from the copper island of Cyprus – but she was also the goddess of love, she was also the morning star and evening star. Now, that was an enormous discovery – when it was discovered that the morning star was the same as the evening star – because you had to watch for a long time to make it. After watching for a long time, people noticed that when there was a morning star there had not been an evening star the previous evening and *vice versa*. Thus it took a long time to identify the morning star and the evening star. Saturn, which moved the slowest, was lead; Mercury was, of course, quicksilver.

These analogous pictures were carried to a much greater

extreme by the Chinese, who had five elements in a sense more practical than the Greek elements. As you might expect, there was water and there was fire, but they had two others, wood and metal, very practical elements to have. Then there was the fifth element which, from the Chinese point of view, was the most important element, earth. The Chinese made the most complicated connections between the elements with everything you could possibly think of, but the most obvious connection was with the colours and the directions: red, fire and the south; green, wood and the east, the damp forest part of the country; black, water and the north; white, iron and the west, where the iron came from; yellow, the earth or the centre, the land of China.

4

Greek Physics

We must now return to the Greeks for the origins of physics. Of course, *we* know what physics is, but they did not. They confused what we would consider biological and theological with what we would call physical. But it was at the beginning of the fifth century BC that these things began to separate out, and Pythagoras is really the key figure in this because his work led in different directions – mathematical and physical. The only part of the mathematical line I propose to follow is that of irrationals that led from the works of Pythagoras, although he did not himself discover it as far as we know. It was the simplest, right-angled triangle which led straight into difficulties. Fig. 26 shows a right-angled triangle with two unit sides and Pythagoras said that the square on the hypotenuse is two units. Therefore, the hypotenuse is what we would now call root 2. But what is root 2? You can measure it, but can you work it out? What number is it? It was one of the followers of Pythagoras, we do not actually know whom, who said that you cannot find a number for it; it is not quite whatever number you think of for it. The Greeks did not have the same notation as we do, so they could not write 1.414 and so on. They were deeply

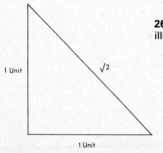

26 A right-angled triangle with two unit sides, illustrating the theorem of Pythagoras

1 Unit

√2

1 Unit

shocked by this situation; it was irrational, unreasonable. That numbers, which should be the most reasonable thing in the world, should not be reducible to whole numbers or their fractions was considered to be a very bad blot on the universe as a whole. This argument about the irrational runs through the whole of the mathematical side of Greek thought and, of course, it is a very profound notion. The existence, beside the integral numbers, of irrational numbers, some roots and others so-called transcendental numbers like π, very much worried the Greek thinkers. The way they escaped from it was not to do as the Babylonians did, but to avoid entirely dealing with numbers and to deal instead with geometrical figures where questions of rationality did not arise. Yet, as Euclid showed, it is possible to carry out rigorous logical deductions from the unspecified length of the lines.

However, as I am not dealing with Greek mathematics, I shall pass that over and consider the physical and philosophical consequences of the work of Pythagoras which led to two quite different developments. After all, what he had done in principle was to reduce the universe to numbers. Now, that could be taken either way: you could say that numbers are the only reality and the universe is a fiction, or you could say that these numbers correspond to real things. But what things? The first line was that of Parmenides of Elea in South Italy, who was a very strict and conservative thinker and maintained that the world as a whole was one perfect sphere because that was the only thing that existed fully in itself and any appearance to the contrary was an illusion. With all this variety in the world, the only thing that could really be relied on is that one is one. Although this seemed to be a rather empty argument, it led to thoughts which were of profound interest to mathematics. Parmenides' pupil, Zeno, put forward some paradoxes that we have not resolved yet. Zeno discussed the arrow that is shot from a bow: it is either in this place or it is not in this place, and if it is moving, where is it in the interval? What is an interval? The whole of the ideas of mathematics, the whole of the ideas of analysis are wrapped up in the paradoxes of Zeno. Some of them are more trivial, such as that of Achilles and the tortoise: the tortoise always managed to keep ahead of Achilles, simply because every time Achilles moved forward, so did the tortoise and Achilles was never able to pass the tortoise. These arguments were used not as mathematical logic, as we would use them, but rather to show that you could not believe what you saw and that, therefore, all appearances were deceptive.

Atoms

The other development, however, led through Leucippus to Democritus, who is always associated with the idea of numbers applied to actual things. The actual things are not things that anyone can see, they are the basic things that exist in the universe. But even before that, there were various other ideas. Heraclitus and Empedocles and others thought the world was full of little seeds that could be turned into anything. But Democritus took a very rigid view and said that there were just hard, unbreakable, *uncuttable* things – *a-tomos* – and that by arranging them in various geometrical figures, you could produce all the various appearances that were seen. The appearances were real because they were made of real things, they were made of atoms. But what about all the rest of the world? Democritus argued that there was simply no rest of the world. The rest of the world was just nothing, emptiness. The world consisted of the atoms and the void. The atoms, of course, did not stand still, they moved around forming new combinations. These are very general ideas; if you try to follow them too far you get into difficulties – and the Greeks certainly did.

This led to a school of philosophy which is always considered a rather dangerous school, a very materialistic school. The most famous philosophers of this school, though much later, in Roman times, were Lucretius and his great master Epicurus. Epicurus applied this view of atomism to humanist ends: the main consideration was what you could actually see and feel. He included many things which we would not consider material at all, such as atoms of thought. For instance, the world was made out of these atoms and you just lived in this present world, the object of life being pleasure. But the pleasure of Epicurus was an extremely refined kind of pleasure – he was called the Garden Philosopher. He spent his time discussing and enjoying a very simple life, removing altogether the idea of duty and gods and so forth, a view which was later refuted by the Stoics of those times. We have, incidentally, very few actual texts. If you go to a library and ask, for example, for Thales's works, you will not find them or even a single statement that could confidently be attributed to him. You can only find second-hand and third-hand accounts of Thales. The same is true for Pythagoras. In fact, none of the people I have mentioned seem to have left any books behind them, although they probably did leave books which have since been destroyed. But sometimes there are quotations from them in other people's books.

By the time we get to the Romans, however, there are a considerable number of books and one of these, preserved for some reason – it is a delightful book to read – is that of a very late Epicurean, Lucretius, who wrote at some length about the whole of the universe. It is in the form of a poem called *Of the nature of things*, which contains a clear exposition of the atomic world. Here are some extracts:

> Material objects are of two kinds, atoms and compounds of atoms. The atoms themselves cannot be swamped by any force, for they are preserved indefinitely by their absolute solidity . . .

> Pay attention, therefore, while I demonstrate in a few lines that there exist certain bodies that are absolutely solid and indestructible, namely those atoms which according to our teaching are the seeds or prime units of things from which the whole universe is built up. . . .

> There are therefore solid bodies . . . And these, as I have just shown can be neither decomposed by blows from without nor invaded and unknit from within nor destroyed by any other form of assault . . . Hence, if the units of matter are solid and without vacuity, as I have shown, they must be everlasting.*

Here it goes into a rather interesting analogy which ties it up with genetics:

> No species is ever changed but each remains so much itself that every kind of bird displays on its body its own specific markings. . . . For, if the atoms could yield in any way to change, there would be no certainty as to what could arise and what could not . . . nor could successive generations so regularly repeat the nature, behaviour, habits and movements of their parents.**

*The Nature of the Universe, Lucretius, Penguin Classic, London, 1951, pp. 42–43.
**Ibid., p. 44.

Here, you see, you have the gene theory of inheritance mixed up with the theory of atoms and, of course, the full essence of that mix-up has only been revealed to us in the last few years with the elucidation of the nature and actions of the molecules of DNA – but we must leave that for the moment as it is not physics.

Lucretius goes on to say that the atoms move and that it is by the atoms moving that you get the variety of things. But now comes the difficulty. The atoms move in empty space and by what might be called the principle of persistent motion or even inertia. At first sight, it would seem that they always have to move in straight lines and fall through space. You must remember that the idea of falling was considered to be an intrinsic or necessary one. The idea that things were suspended and were equally liable to fall in any direction could not be envisaged at all. Atoms could not stay just in space: they had to be always falling. If they were always falling, how did they meet each other to make all these combinations? There, of course, some special help was needed, and this is the weakness of the whole argument:

the atoms swerve a little – but only a little, or we shall be caught imagining slantwise movements, and the facts will prove us wrong . . . The sum of things cannot be changed by any force. For there is no place into which any kind of matter might escape out of the universe or out of which some newly risen force could break into the universe and transform the whole nature of things and reverse their movements.*

You notice the basic ideas there? These ideas of atoms go right through to the present day and, even in their first form, they represent good observation and fundamental physics so far as it goes. The argument is that things do not change, that the properties of things are always the same and, therefore, that atoms cannot change. Until this century, in my own lifetime, this view held. The two main things that have been left out in this theory are atomic fission and radio-activity, on the one hand, and mutations on the other. Lucretius thought that the birds always had the same patterns on their feathers. The idea of the appearance of a new kind of bird with new patterns on its feathers, the idea that ducks and swans

*Ibid., pp. 67, 69.

and birds of that kind had a common ancestor had not yet arisen. You took a certain small time-section of the world and within that time-section nothing changed, everything went through a cycle, everything had a peak, and it was simply a matter of re-arrangement, in a practical way, of the same things. This notion goes through the whole of the history of science from that time to Darwin.

But this atomic theory was not on the main lines of thought at the time, as you will see. For various reasons, mostly religious and social, it was considered a very dangerous hypothesis. Curiously, nevertheless, it was never forgotten. It was pushed on one side and, instead, the development of science proceeded much more on the older lines of assuming that the essence of things was not atomic but continuous. This was the view taken by the greatest of the Greek scientists, Aristotle.

The tyrants

I must say a little here about Greek history. Most of you will know much of it, but in case any do not, it is very relevant at this stage. The early philosphers I have talked about lived in small towns and ports all over the boundaries of the Greek world. These small towns were at first ruled by the best families and landed aristocrats. Later, some of the merchants in the towns made so much money that when the aristocrats were rude to them, they were able to drive out the aristocrats or kill them and set themselves up as what the Greeks call tyrants. There was an age of tyrants in the sixth century BC: there was a tyrant in Athens, Pisistratus; all the Ionian philosophers lived in the time in one or another of the tyrants' cities. Then in Athens particularly, but also in other places, there was a movement against the tyrants. It did not come even mainly from the old aristocratic families, it came from the smaller tradesmen, from what were called the people – not exactly what we would call the people because some half to one third of them were slaves. The smallholders near the town and the small traders and free workers in the town revolted and killed or drove off the tyrants and set up a government of the people, a democracy. Sometimes the tyrants would go to the king of Persia – Pisistratus's family did that – and invite the Persians to get rid of this democracy which was endangering the empire of Persia by encouraging the Greek cities in Ionia to rebel. It was obviously not tolerable to have this situation in Athens, so close to the coast of Persia, and so a great expedition against Athens was made by the Persian Emperor, Xerxes, with Pisistratus helping to show the way. The Persians took Athens and burnt most of it. But they could not get much further. Their fleet was, in turn, sunk by the Athenians at Salamis and the Persians retired, so to speak, in a huff: although beaten in a further battle they were never *driven* out of Greece but they left it for good.

After that there was a new set-up taking the form of a number of democracies, mostly under Athens, the so-called Athenian empire. It was done in this simple way: the Athenians provided the armaments and the ships – they had very good silver mines, so they had lots of money – and they were considered to be, or they considered themselves to be, the leaders of a great confederacy. At this time, the Persians were still a potential threat with the Persian king still ruling in Persia; they stirred up a war between the Athenians and the much more primitive people, the Spartans, who were just plain tribesmen devoted entirely to warfare. They were real Junkers, that is, they were landowners, all the work being done, not by slaves but by serfs, the Helots, who could not be sold as they belonged to the land. In the end, after a very long war in which most of the things in Greece that were useful had been destroyed, the Spartans won – with Persian gold.

The Academy

Now, all this is relevant to science because it was precisely at this time that philosophy in mainland Greece turned away from physical science and towards moral and political issues. The chief philosopher in Greece, Socrates, we know would have nothing to do with physical science at all. Socrates considered that people wasted their time in studying the natural sciences because such studies had not helped the soul in any way. People should have studied the moral sciences. The chief pupil of Socrates, Plato, went further. Plato had a very important part to play in science – not in the experimental science I am talking about here but in theoretical physics. Plato was a Pythagorean, he believed in the enormous importance of mathematics. He set up a school which was called the Academy and he had written above the door: 'Let no one ignorant of mathematics enter here.' He believed that if you understood mathematics well enough, not only would you understand the universe but you would understand how to behave. He was also the supreme writer of Greek prose and he had an enormous effect on intellectuals, particularly later on in the bad times after the fall of the Roman Empire. But he limited the world to what could be expressed in mathematics. He had some rather wrong mathematical ideas, but he did stimulate its study. Eudoxus, his mathematical follower, so to speak, actually devised a kind of machine which explained the apparent motion of the sun and stars around the earth, their rising and setting.

Aristotle

One of Plato's more famous pupils was Aristotle. By this time the power of the Greek cities had declined and a new power arisen, that of the barbarian Macedonians, semi-Greeks to the north. The Macedonians had not reached this stage of democracy, being still at that of kings and nobles. Philip of Macedon conquered the cities of Greece or, rather, he had what you might call a 'good neighbour' policy for Greece. He made them all join a league, a league with no power of any kind, all the power being in the hands of Philip. Philip had a son of whom everyone has heard, Alexander, and Aristotle, pupil of Plato, was his teacher. Alexander must have been a rather difficult person to teach because it is alleged that Aristotle said to him: 'There is no royal road to mathematics, you have to work at it.' (Royal roads were the main roads in the Persian empire.) There is no evidence that Aristotle got any mathematics or science into Alexander but it did give Aristotle enormous opportunities for learning more about them.

Subsequently, Aristotle made a deep impression on all the science of his times and, in fact, it may be said that the great movement of modern science, the great Scientific Revolution in the sixteenth century, which we will come to later, was a revolt against Aristotle's teaching in all respects. Only recently has it been understood that Aristotle as a scientist was not a physicist; he was almost as much the opposite of a physicist as you could have been then – he was a biologist. He treated the whole of the universe as if it were a kind of living thing. Aristotle's physics particularly – why things fall and why they move – was based on observations of animals. If you observe animals, you see that whenever they do anything it is for a purpose. It is to get something to eat or to avoid something that is chasing them. And Aristotle interpreted the physical world in the image of the animal world. He made marvellously accurate descriptions of the animal world, some of which have not been improved on until quite recently. But when he came to physics, he did not introduce anything new; rather, he went back to much cruder ideas that had been forgotten by the more sophisticated scientists. He had not much use for mathematics and his picture of the solar system was very crude. He went back to the idea of the earth surrounded by these layers of elements – earth, water and so forth. The things moving in the sky were supposed to be, as it were, pins stuck into an arrangement of rather complicated revolving spheres, a very mechanical kind of universe which became more and more rigid as time went on. Aristotle did not really mean that they were rigid spheres, but in the Middle Ages it was believed that all these celestial bodies were stuck in some super transparent glass spheres – their motions had to be transferred – and as there was an awful lot of them, they had to be carefully geared one inside the other. I think there were about fifty-three spheres each one of which operated the ones inside, and, of course, to keep this terrific machine going you had to apply power, and the power was applied to the outer sphere which was the prime mover, moving everything around inside. This kind of astronomy did not really help the development of technical science but it did enormously improve the prestige of mathematics.

What Aristotle did, too, was to set up a school; his school had a different name – it was a Lyceum. It was not like Plato's school. Plato's Academy was based on promising young men in Greek politics and they could discuss things like truth, political justice, the use of force – the kind of things you would still learn if you went to Oxford. It was the real 'Greats'. But Aristotle set up as well what you might call a research institute – the Lyceum. The people there had to look at minerals, had to make observations of winds and similar phenomena and there were also biological and social studies. His research school collected – it must have been a terrible job – the constitutions of 158 big cities, their methods of voting, *etc.* in order to make a comparative study.

Now, all this activity was raised to a much higher level in the next stage of history. Alexander was extremely successful in warfare. He used the latest methods of Greek technology and he just walked through the Persian Empire, chasing the king of Persia and contriving to get him killed. He took over the whole Persian Empire. After this he marched right through Egypt – and, in fact, all over the world then known. Practically the whole area shown on Map 2a was Alexander's empire.

Unfortunately, he could not stand the drink and the climate of Babylon – and I am not surprised – and he died at the age of thirty-three, complaining that he had no more worlds to conquer. He had conquered practically all the civilised part of the world. However, the conquest did not last for more than a year, in the sense that Alexander's generals started quarrelling with each other at once and after long and tedious national warfare, sorted themselves out into four kingdoms (Map 2b). General Seleucus's kingdom formed the entire area ruled over by the Antiochid dynasty which for some time controlled India. India, however, was the first to break free from the Greek rulers and, then, copying the Greek devices, set up an enormous empire – the empire of Asoka – in which the Greek influence very largely disappeared. But the most important general was Ptolemy, who took Egypt. Most of modern science comes from Egypt, not the Egypt of the Pharaohs but the Egypt of Alexandria.

The Museum

In Alexandria, the pupils of Aristotle, particularly his favourite pupil Theophrastus, set up a new kind of institution, a really learned institution which they called a *Museum*, for the Muses. This was an all-in institution; there had never been anything quite like it. Not only were there all the dramatic critics, poets, painters and other such people in one wing of the Museum,

Map 2 The Conquests of Alexander

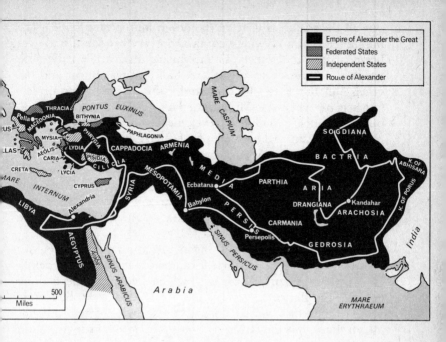

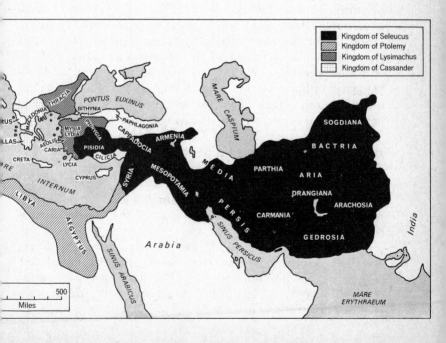

but all the scientists were in the other, and joining both was a magnificent library which contained everything that had ever been written in Greece. Later, during the rather ridiculous episode of Caesar and Cleopatra, half of it was burnt and the other half left to go to rack and ruin. It was more or less finished off when the Mohammedans conquered Alexandria, but in its time it had been extremely well used, as were the research institutes. There people received salaries, whereas the philosophers I have been talking about lived on fees, on what they could pick up, and on patrons of one sort or another. When you worked in the Museum at Alexandria, that was considered a job. Once you were accepted there, you were there more or less for life, and as a result the sciences were really developed.

Most of the mechanical things I talked about in a previous chapter were developed in Alexandria. The Ptolemies really exploited the country in a way which had never been done before. The Pharaohs, in their old-fashioned style, had done some things: they would take what the Nile offered and not ask for more. But the Nile only flooded up to a certain level and there was a great deal of flat land above that – it is still there, of course. There would be a field on one side of which the crops would be planted, and a small irrigation ditch would be put in – and the sandy waste would be green. Cultivation went as far as and no further than where water was available. But the Alexandrians, by a government system, introduced all kinds of water-lifting devices – shadoofs, water wheels, Archimedean screws; they developed gearing and a great deal of general engineering. They discovered, among other things, little steam engines; Hero of Alexandria wrote about them in the first century AD but he probably did not invent them. The Alexandrians also had automatic devices for opening temple doors and various technical tricks of one sort or another. Yet they also used a great deal of mathematics and a great deal of real physics – Strato worked on most of the laws of hydrodynamics.

It was, however, in an offshoot of Alexandria, Syracuse, also under a series of Greek tyrants, where the most remarkable of these natural philosophers lived. Now, as I have said, moral philosophy more or less came to an end with Alexander's conquests or, rather, it turned into a different kind of philosophy which does not concern us here; it became concerned almost entirely with ethics. You had the School of the Stoics, founded by Zeno, who said that duty was the only thing that mattered and that if a person was virtuous and did his duty, virtue would be its own reward, and everyone had responsibility for the whole universe. Only the very 'top' people were Stoics, like the Confucians in China. It was the religion or philosophy of the best people. The slightly less careful best people were Epicureans and most of the ordinary people took to various other salvation religions.

Archimedes

The scientists, however, kept apart from this. They studied particularly geometry and mathematics, of which I am not speaking here, but they also developed quite a lot of what we call physics. I could mention pneumatics and hydrostatics, the main development of which came from Syracuse and particularly from Archimedes. His works, which have been well preserved and passed on, dealt very largely with the subject of mechanics, being rational justifications for the actual technological developments already referred to. Most were based on the idea of the lever, the way in which forces passing through different points could balance each other: 'Give me a fulcrum to lean on and I could move the world.' With that he also studied the balance of actual bodies of different shapes and developed the idea of centre of gravity. All this was highly theoretical. We know, however, from manuscripts which have been preserved, that Archimedes also made actual models to see whether his theories corresponded with facts. His most famous achievement is the development of hydrostatics leading to his celebrated idea in his bath, that the force on a floating body is equal to the volume of the weight of the water displaced, that led to the whole practical measurement of specific gravity.

Archimedes by no means monopolised the Greek science of the Hellenistic period. Remarkable developments were made in astronomy. I referred earlier to the rather crude astronomy of Aristotle, but in the Museum of Alexandria, his followers had largely abandoned it; their work was too refined to be popular and when the works of Aristotle were again studied in the Middle Ages, his views were preserved and not those of his critics.

The improvement in astronomy, actual computational astronomy, especially in Egypt, with close approach to the Babylonian sources, was enormous. Hipparchus made star catalogues, eclipses were studied, and the first models of the solar system were made, models which involved difficulties to which I will return later. The simple picture of the celestial system is that you have the earth at the centre and celestial bodies going round it: first the moon, which goes round it once a month, next the sun, which goes round it once a year, and then there are the wandering stars, the planets, which behave in a rather strange way. They all go round almost every year, sometimes in a little less, and appear to move in a rather peculiar way, never moving far from the sun. There are two planets which always go with the sun and never get far from it, Mercury and Venus. They rotate, but they only rotate with the sun, which is why they sometimes go round a bit faster than the sun and sometimes a bit slower. Mars follows a very queer course, it kind of loops the loop all the way round. Jupiter loops the loop too, but in much smaller loops – twelve of them – and Saturn, which goes slowest of all, has thirty-two loops, each loop taking precisely one year – it takes over thirty years to get round the heavens. All the motions are subsidiary to the daily motions of the whole of the heavens round the earth, once every twenty-four hours.

The simplest model of that is one where all the planets would be on eccentric paths. The planet goes round on a perfect circle, about a point, not necessarily at the centre of the earth, but somewhere on a line joining the centre of the earth with the sun. These eccentrics are perfectly good to account for the observations. If you add enough eccentrics, you see the process becomes purely mathematical; any orbit can be approximated. If you want to produce an orbit very accurately you have to have rather a lot of eccentrics; but if you are an astronomer you do not mind because it only means a little more calculation.

By the use of these orbits and eccentrics, which were worked out in Alexandria slightly later by Ptolemy – Ptolemy the Astronomer, not one of the royal Ptolemies – you can produce tables which give you pretty good predictions of all the planetary positions for many years ahead. They were not the first tables, these had been made by the Babylonians, but they

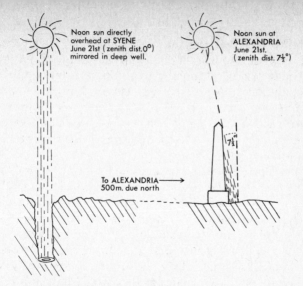

27 Eratosthenes' method of measuring the Earth. He noted that at Syene the sun was directly overhead on Midsummer's Day, whilst at Alexandria the sun's rays were 7½° from the vertical, a value calculated from the length of a shadow cast by a column of known height. From this, and knowing the distance between the two places, he calculated the circumference of the Earth to be 250,000 stades (about 24,000 miles)

were to become, with a series of improvements, the basic guidelines of the motions of the stars and the planets for future centuries and to lead to the great discoveries of the Renaissance. The importance of this work was heightened by the belief that if the planetary positions could be known, the fates of man and nations could be determined, as astrology had become the mother of astronomy.

A round earth

One consequence of this was the confirmation of an earlier, Pythagorean theory – that the earth was round. This was particularly easy in Egypt because the peculiar thing about Egypt was this: it is at about latitude 30° and the tropic is at latitude 23½°. Now, people knew that on Midsummer's Day the wells in Syene were lit up by the sun, which was considered to be a great marvel, but this for only one or two days in the summer solstice, or summer 'still stand,' when the sun stood vertically above them in that place at noon. At the time, the sun was 7½° from the vertical at Alexandria (Fig. 27). One possibility was that the sun was somewhere fairly close, but when they came to work it out – and they did work it out – they could not find any solution based on this possibility. Obviously, if the sun just stood there fairly close, even on a plateau, it would be at an angle; but if you tried to work it out to actual degrees – which they had by then – you just could not make it fit. The sun had to be very far away and its rays at Syene and Alexandria parallel.

So the only possible explanation for the $7\frac{1}{2}°$ was that the earth really must be round and this enabled them to divide the earth into zones – zones of latitude – those in which the sun at some time of the year was vertical. The tropic zone was not so called because it was hot but because it is where the sun is on the top point of its journey, or turns round, 'tropos'. At Syene, on Midsummer's Day, the sun stands absolutely vertical and then it goes down again; as they would have put it, that is where the sun turns round. South of Syene, it was tropical.

Then they gradually learned that there was a zone up in the north where the sun did not rise at all, that was in the region of the domination of what we call the Pole Star, a star in the Great Bear; that was called the Arctic or Bear Zone, and in between was ordinary temperate country. No one could live in the Arctic Zone or continuously in the Tropic Zone. They thought it became hotter and hotter as you went further south – it was pretty hot at Syene – and, therefore, it was impossible for anyone to go further south and they might as well give up trying, although some people, not knowing enough astronomy, had actually done it, and some of the Carthaginians had sailed round Africa.

The point of this understanding is that it led to the picture of a really spherical earth and, soon after that, to the possibility of this solid earth turning round. In fact, it did, but this was considered at the time of the Greeks and for many centuries afterwards an absolutely outrageous and impious thought.

We have in the work of Archimedes a complete account of this, which otherwise would have been quite lost. At the time, Archimedes was dealing with something quite different. He was in the service of King Hieron of Syracuse and, afterwards, of another tyrant who was less lucky. Archimedes was concerned just then with concepts of large numbers. He wrote a little treatise which he called 'The Sand Reckoner' in which he considered a new way to count how many grains of sand there are. He said:

There are some . . . who think that the number of the sand is infinite in multitude; and I mean by the sand not only that which exists about Syracuse and the rest of Sicily, but also that which is found in every region whether inhabited or uninhabited. Again there are some who, without regarding it as infinite, yet think that no number has been named which is great enough to exceed its multitude. And it is clear that they who hold this view, if they imagined a mass made up of sand in other respects as large as the mass of the earth, including in it all the seas and the hollows of the earth filled up to a height equal to that of the highest of mountains, would be many times further still from recognising that any number could be expressed which exceeded the multitude of the sand so taken. But I will try to show you by means of geometrical proofs, which you will be able to follow, that, of the numbers named by me . . . some exceed not only the number of the mass of sand equal in magnitude to the earth filled up in the way described, but also that of a mass equal in magnitude to the universe.

Now you are aware that 'universe' is the name given by most astronomers to the sphere whose centre is the earth and whose radius is equal to the straight line between the centre of the sun and the centre of the earth. This is the common account, as you have heard from astronomers. Aristarchus of Samos brought out a book consisting of some hypotheses, in which the premises lead to the result that the universe is many times greater than that now so called. His hypotheses are that the fixed stars and the sun remain unmoved, that the earth revolves about the sun in the circumference of a circle, the sun lying in the middle of the orbit, and that the sphere of the fixed stars, situated above the same centre of the sun, is so great that the circle in which he supposes the earth to revolve bears such a proportion from the distance of the fixed stars as the centre of the sphere bears to its surface.

Now that is a difficult one for any mathematician!

Now it is easy to see that this is impossible; for, since the centre of the sphere has no magnitude, we cannot conceive it to bear any ratio whatever to the surface of the sphere. We must, however, take Aristarchus to mean this; since we conceive the earth to be, as it were, the centre of the universe, the ratio which the earth bears to what he describes as the 'universe' is the same as the ratio which the sphere containing the circle in which he supposes the earth to revolve bears to the sphere of the fixed stars.

Then Archimedes goes on to say that if a sphere, even one as great as Aristarchus supposes that of the fixed stars to be, were made up of sand, he could still find a number which expresses the number of grains of sand in it, provided the following assumptions were made: The perimeter of the earth is about 3,000,000 stadia (stadium = about 500 feet) and not greater; the diameter of the earth is greater than the diameter of the moon, and the diameter of the sun is greater than the diameter of the earth. Well, this, you see, is shocking, because, only 150 years earlier Anaxagoras had been expelled from Athens for saying that the sun was as big as the Peloponnese, now it is said to be bigger than the earth. However, Archimedes had the figures a bit wrong, because he then says that the diameter of the sun is about thirty times the diameter of the moon and not greater. These celestial distances, you know, are very difficult to measure!

The point I want to make, however, is that the Alexandrians had a very sophisticated understanding of the essence of the physical world. So we have now reached the stage of two basic theories about it, which were to be revived later on: one that the earth goes round the sun; and the other that everything is made out of atoms. The Alexandrians had elaborate methods of calculation, rather weak on the numerical side, but very strong on the geometrical side. Incidentally, all the geometrical forms, all the shapes, particularly of the conic sections, were well studied as pure curves. They were first studied as shadows: take an object, a circular ring, throw light on it from various directions on to a board and you get projections which are ellipses and hyperbolas and so forth. Appollonius made a complete study of these curves – and there he had the whole basis for a geometrical understanding of the universe.

In 212 BC, all this knowledge came to a stop, not for scientific reasons, but as a result of political events one of which even stopped Archimedes himself. Syracuse, in the eastern part of Sicily, was then a kind of bone of contention: the island's eastern part was Greek and the western part was Carthaginian. The Romans were at war with the Carthaginians and the Romans were a tough lot and rather uneducated. They borrowed some ideas from the Greeks and used them practically but never really appreciated them. Greece was rather weak and took the wrong side in the war, allying herself with the

Carthaginians. Syracuse was besieged and Archimedes was called in as scientific adviser, using mirrors to burn up ships and other kinds of devices to outwit the Romans, but to no avail. The Romans took the town and by accident killed Archimedes. It was the act of a soldier whom Archimedes was berating, as we are told in Plutarch. The legend is that Archimedes was working out a geometrical problem and the soldier, rushing through the town bent on general loot, came across him and told him to get out of the way. Archimedes said to him: 'Don't disturb me, I am trying to work out this problem', whereupon the soldier just killed him. The Roman general, who was a man of culture, was very distressed, but it was done and that was the end of the development of the theories of Archimedes.

Nevertheless, things did not quite end like that. A number of good works were produced in other fields, particularly in medicine, during Roman times. The goal, so to speak, had disappeared from physics but it lingered for a while in mathematics. The last of the Greek mathematicians, curiously enough, was a woman, Hypatia; she became involved with politics but on the pagan side, not the Christian, and she was killed by a mob of Christians bent on destroying the last temples in Alexandria. After that there was no original science for a long time.

The legacy of the classical world

I shall now try to tell you something about what was left and how it was rediscovered, because there is no doubt that from about 100 AD until almost 1200, and some people would put it as late as 1540, the sciences did not even reach the Greek standard. In 1540, the works of Archimedes, which were known, were translated for the first time into Latin and they were read by competent mathematicians. That led, among other things, to the great revolution that gave us modern science. The Greeks did not live to make that revolution but they bequeathed to us an enormous mass of material which we now take for granted and out of which modern science was to be born.

After that came what might be described as the frills. The basic things, the machines, the methods of agriculture and so forth, remained, everything that was not on too large a scale. If you go to Rome, you will see all around it magnificent aqueducts, built in classical times, which have been damaged at various times since and which, of course, are now of no use as aqueducts. But Rome has not as good a water supply now as she had then. Small-scale organisation was just as good in the Middle Ages as it was in classical times. We are apt to be so dazzled by the intellectual and artistic brilliance of the Greeks that it is hard to realise that their knowledge and skills affected far more the appearances than the practical and material sides of life. The beauty of Greek cities, temples, statues and vases, the refinement of their logic, mathematics and philosophy, blind us to the fact that the way of life for most people in civilised countries was, at the fall of the Roman empire, much what it had been 2,000 years before, when the old Bronze Age civilisation collapsed. Agriculture, food, clothes and houses were not notably improved. There were slight improvements in irrigation and roadmaking. The science of the Greeks found little application in the new styles in monumental architecture and town planning.

This was not surprising. In the first place, science was developed by well-to-do citizens, not for social purposes, which they despised. In the second, even with the best will in the world, the science they acquired was far too limited and qualitative to be of much practical use. Greek mathematics, elegant and complete as it was, could be applied to few practical purposes because of the lack of experimental physics and practical mechanics. Perhaps this is not quite true; there were slightly more accurate mechanics than we believed up until a few years ago, but the finest efforts, like clocks and so forth, were mostly destroyed. The great nursery of applied astronomy, the art of the navigator, hardly existed for lack of ships or incentive to sail in a trackless ocean. In the Mediterranean they did not really need any navigation; you just sailed a course and you were bound to hit one side or the other – you could not really get lost in the Mediterranean. The other natural sciences were hardly more than discursive catalogues.

The full possibilities of classical culture could not be realised in the framework of the civilisation that gave it birth. They were blocked at every turn by the social and economic limitations inherent in a slave-owning plutocracy. The real contribution of Greek science was to be seen in the future, though it could only be made in so far as the germinal element of classical culture could be preserved and transmitted – and that is what happened. Indeed, practically the most important thing to be salvaged from the classical age was the *idea* of natural science. Yet the belief persisted through legend that the Ancients, through deep study, had acquired the knowledge of nature which enabled them to control it. The legend had it that Alexander, instructed by Aristotle, had a submarine and could fly through the air in an eagle-powered chariot; there were plenty of pictures of these.

Of all the elements of classical culture, it was science, particularly astronomy, that proved in fact the most lasting. Much of the other sciences of the Ancients was preserved in books; but because astronomy was needed to chart the planets, if only for astrological predictions, it had to be handed on and practised. There is no break in the practice of astronomy, that is the most important thing. Observations were carried out year by year from about 2000 BC to the present time continuously. If there was a break in any given place, if one observatory was knocked out, there was always another one; they were never all broken up at the same time.

People were very soon converted to astronomy because, as I have said, they needed it for astrology. The Tartars were generally considered to be very tough, burning everything in their path, but within fifty years of the Tartar regime in Central Asia the Il Khan of Tartary set up not only the finest observatory that had ever been there, but made the best astronomical tables of the time.

So there was a continuity, but the tragedy of it is, why did we have to wait 2,000 years for its rebirth? That will be the subject of my next chapter.

5

Medieval Physics

The next period with which I shall be dealing is strictly an intermediate one but, as we shall see in hindsight, it is vital to the development of science for preparing the Scientific Revolution. In fact, it did far more than that directly. What happened at the end of the classical age, in the break-up of the Roman empire and even before that, was the disappearance, though not permanently, of one particular cultural achievement – or rather its going down into little roots and seeds that would grow later. Science was no longer taught and, in the West, we went into a period which is generally referred to as the Dark Ages. This, however, is because we have a rather one-sided view of it. If you ask people what was happening in the fifth, sixth or seventh centuries, very few people, even professional historians in the West, will be able to tell you. What was apparently happening in Europe was what is called the folk wandering: the Franks were coming into France, the Angles were coming into England, the Burgundians, the Lombards and various other tribes were wandering about all over Europe. The Vandals were going into Africa, and the Huns were coming over into the whole of Eastern Europe. Naturally, in the process of this mass migration, much of the learning, much of the civilisation, disappeared, particularly its large-scale operations.

What was more important in the West was the disappearance of cities, in some cases completely, more so in England than anywhere else. For instance, a very fine and civilised capital city, with theatres and baths and many other amenities at St Albans, the civil capital of Britain, was left completely deserted. So much so, that the new city which grew up in the tenth and eleventh centuries was built entirely out of the bricks of the old one. Because the Saxons did not know how to make bricks, they used the Roman bricks where they could find them. But this disappearance of cities was a very uneven process. While it applied to outlying places like Britain and to a minor extent to Northern France, it did not apply at all in Italy where the cities lived on. It is true they changed in character: many of the big buildings in Rome decayed, but there were enough left to impress people for many a long day afterwards. The Forum was turned into a pasture for cattle, the marble statues were burnt for lime, yet there was some degree of continuity of urban government.

All this talk of a Dark Age gives as I have said a one-sided view. If you ask what happened in the fifth century in some other parts of the world, you will see that civilisation did not decay there at all. Rome was in a bad way, but there were no buildings destroyed at all in Constantinople. Not only were none destroyed, but the finest building in the whole of Europe and one of the finest in the world was built there during that period – the church of St Sophia, which is still the most daring dome construction in the world, much bigger than St Paul's or St Peter's. Similarly, at Alexandria, Antioch and in other parts of the Near East and further East still, in India and in China, enormous cities were built. The capital city of Sian, in China, for instance, which flourished at that time, covered an area not very different from that of present-day London and was much better planned, with streets, markets, gardens and so forth, although now it is very largely deserted. And many other such cities were built. Thus, there was no break in the continuity of civilisation as a whole, but there was a break in the part of it that we recognise and, therefore, it seems to us much more of a break than it really was.

The spread of Hellenism

Here, I should emphasise that the Greek ideas were spread enormously widely by Alexander and his successors. The Hellenistic world picture, which was itself a mixture of sciences from Greece, Egypt, India and Babylonia, was spread right over what might be called the old world (Map 3). It did not spread into America at this stage, but from Peking to Fez you had people who could argue the same way and had the same mathematics and astronomy – which was the universal

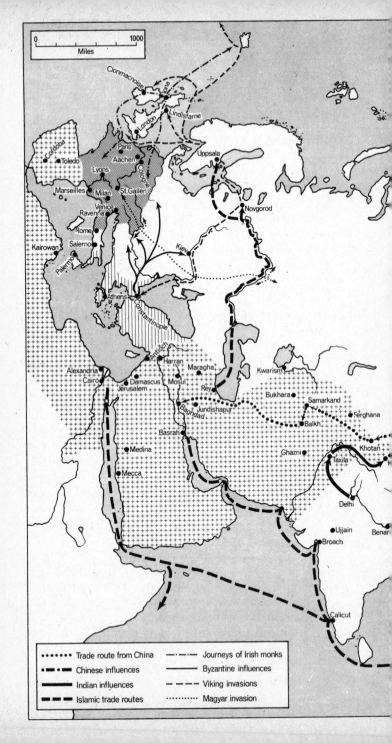

0		1000
	Miles	

Clonmacnoise

London
Lindisfarne

Uppsala

Córdoba
Toledo
Paris
Aachen
Lyons
Cologne
St. Gallen
Milan
Marseilles
Venice
Ravenna
Rome
Kairowan
Salerno
Palermo

Novgorod

Kiev

Athens
Constantinople

Antioch
Harran
Maragha
Kwarism
Alexandria
Cairo
Damascus
Mosul
Rey
Jerusalem
Bukhara
Samarkand
Jundishapur
Ferghana
Baghdad
Balkh
Basrah
Medina
Ghazni
Khotan
Taxila
Mecca
Delhi
Ujjain
Benar
Broach

Calicut

•••••• Trade route from China	–·–·– Journeys of Irish monks
–··–··– Chinese influences	———— Byzantine influences
▬▬▬ Indian influences	– – – – Viking invasions
▬▬▬ Islamic trade routes	········· Magyar invasion

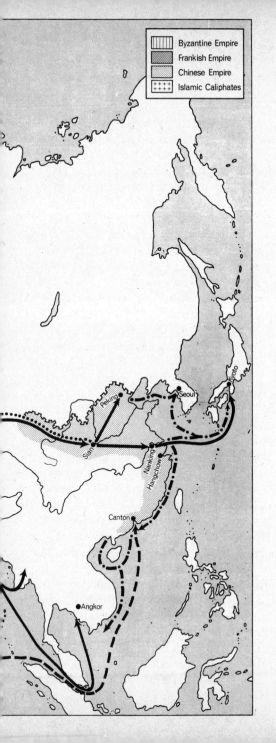

Map 3 The world
in transition

science – and the same geography, quite sophisticated geography with latitudes and longitudes and so forth. You might expect that when this all came back into use, it came from the places that had not been destroyed to the places that had; but as we are interested in what happened in our part of the world, we do not take the occurrences in the rest of the world sufficiently into account. It is only recently that one has been able to understand the process by which the overall world of science of, say, the fourth or fifth centuries AD was saved and transmitted and grew up again in a slightly different form in Western Europe.

The first stage of this process occurred essentially in Syria. By this time Western Europe and the Middle East had been converted to Christianity but, although converted to Christianity, they had not quite been converted to Greek Hellenism. The language of this empire was Greek at this time, not Roman, for the Greeks had just made a kind of comeback, with the shifting of the empire's capital, owing to the Western barbarians, from Rome to Constantinople. This gave the Greeks an enormous advantage. But the Greeks were not very popular in many places in Mesopotamia and Syria because their inhabitants felt that they themselves had a culture which was certainly older and probably just as good as the Greeks.

The first break-away was in language. In the first stage a number of Christians who had broken away from orthodox Christianity – I do not want to go into the heresies, although one of these heresies was very important to physics – had set up the so-called Nestorian Church in Syria. They had not only translated the Bible and the Scriptures into Syriac but also the works of most of the Greek philosophers – Aristotle, Plato, etc. These Nestorians travelled all over the world and, in particular, went to China and North India where they stayed right up to the eighteenth century. And so the first spread was due to the translated knowledge of the Syrians.

Then in the seventh century, there appeared another group, the Arabs, who had extraordinarily little space in history before then except as traders on the Red Sea route, which was the main route between Rome and the west coast of India from where most of the trade came. The great empires that I mentioned earlier – the Graeco-Roman and the Persian – had fought each other more or less to a standstill and at that point the Arabs started their great *jehad* – I must not call it a crusade for it is deemed to be the opposite – a holy war on the whole of the Graeco-Roman and Persian empires. They were extraordinarily successful, and very quickly so. When you get the whole of an area reaching all the way from Spain on one side to Pakistan on the other, all converted to Mohammedanism within about fifty years, you realise that there must have been something behind it.

What was behind it was that everybody, in all those countries, was completely sickened by the Greek and Roman imperial rule and the corresponding Persian rule of the Sassanians. These systems bore very hard on the people and were themselves really responsible for the barbarisation, because taxation was so heavy that people left the towns and went into the country. Mohammedan rule was much simpler and fairer – at any rate, that was what people thought – so they went over to it most willingly. However, Arabian culture had not been on a very high or sophisticated level, and the result was that when they took over all this area they did not administer it in their old way. The administration in Egypt and Syria was carried on in Greek for about 300 years after the Mohammedans took over. This produced another of the phenomena which had occurred at the beginning of the Greek period; there grew up an interest in science which could be separated from the incrustation of all religious beliefs – in this case Christian beliefs – which had been attached to it.

Previously, the monks, because they associated philosophy with all the other things they did not like about Greek and Roman rule, had thrown out the philosophy. They had stoned the philosophers, they had closed the schools. In 527 AD Justinian had closed the schools of philosophy of Athens, which had been going since Plato's time. But when the schools were closed, the scholars were not persecuted in the usual way, they were not imprisoned: in this case, they simply went over to the Persian court where a new school was set up for them in the town of Jundishapur where they continued their work. So there was no break in the continuity, only a break in the locality.

Very soon after that, in Jundishapur itself, the next stage in the process began. First, all the Syriac translations of Greek works were translated into Arabic, and, as by that time some of the Arabs had learned enough Greek themselves, they translated a great deal more. So the main body of Greek knowledge had by that time become Arab knowledge and the wider extent of Arab knowledge added to this meant that now there was a much greater scope for development than in Greek times. There were scholars, and plenty of them, who regularly travelled all the way from Morocco to Peking and back again, or across into Central Africa, or up to the north, rather oddly, through Russia into Scandinavia. We know this because of the hoards of Arabic coins which are found there, and the earliest descriptions of Russia that we have are made by Arabs. So a geographical extension of science, which was much greater even than the extension which had been made by Alexander, was one of the immediate fruits of the Arab conquest.

Islamic science

The Arab period of science can be said to date not quite from Mohammed but from the eighth century to the twelfth century. In all the places where it spread, it was not merely science, it was organised science. It was organised by the foundation of something to which we are now quite accustomed, but which was new in the world of those days, universities. Strictly speaking, there were no universities in Greece at all: there were schools. A philosopher would have a school and a few men would come and listen to his lectures, but it was essentially a private affair. The nearest thing you had to a university was the Museum, but the Museum was not, strictly, a teaching institute at all, it was a research institute. The idea of a teaching institute arose out of the school attached to the mosques, the Madrasah, and there teachers would come and teach all subjects to any students who came there to learn.

There is a very instructive story about one of the greatest of the Mohammedan scholars, Al-Ghazzali, who lived in the eleventh century. He had been to the university of Ray in Persia and, in his four years in the university, he had studied all the courses – philosophy, metaphysics, mathematics – and everything that could be learnt there. He was on his way back to his native town, hoping to make a career for himself, and, being a poor student, he attached himself – as an individual traveller would have had to do – to a caravan. As they were travelling along, a set of Bedouins attacked the caravan and robbed them of everything. Al-Ghazzali had kept all his notes from the courses in a little leather bag which was about all he had, and they took that too. So he went to the Bedouin chief, caught his stirrups and begged him to return this bag, saying that it was of no use to the Bedouins who could not read and that it was the fruit of his four-year learning at the university. The Bedouin chief threw the bag at him and said, 'I thought you went to the university to learn, not to take notes'. Al-Ghazzali was very struck with this and went back to the university for a further four years, taking no notes at all but really thinking about it all to such good purpose that he became the leading Mohammedan philosopher.

So by now, institutionally, you have the foundation of the universities, and the interesting thing is that many of the customs we have now are taken directly from those practised in the Mohammedan universities – for instance, the cap and gown, which was referred to then as the robe of honour and was given to students after matriculating. Four hundred years later, when European universities arose, their founders thought – well, these caps and gowns are part of the university, we must have them too. Similarly, they have very nice caps and gowns in Africa now and many other places.

But not only were there universities, there were scientific societies. They had rather curious names from our point of view. One of the most famous was called 'The Brethren of Purity', which was a chemical society and studied the methods that were supposed to provide the elixir of youth and the stone for the transformation of base metal into gold and so forth. Yet, no one knows to this day whether that was what they were really doing or whether they were not using chemistry as an analogue for moral virtue – the two were curiously mixed. Physics, somehow or other, was never quite so mixed with morality as was chemistry.

Now let us look at what they actually did. One of the later Mohammedans said: 'Our duty is to examine the works of the Ancients (the Greeks) and to improve them in any place in which they can be improved.' This was their attitude: that there was a terrific body of knowledge of the Ancients which ought to be known and ought to be examined to see whether it could be interpreted, tidied up and improved. And this they did.

First of all, they dealt with the primary subject, which I have already discussed and will again, astronomy. They carried on very accurate astronomical observations; they set up a number of observatories with bigger and bigger apparatus – some of the biggest that we have still are at Jaipur in India, which came rather late, and also in Delhi and parts of China. These were all huge ecliptical observatories, as it were, glorified sundials, but being made bigger and bigger so that they gave greater accuracy. I once happened, by luck, to be in Jaipur at noon on the day of the winter solstice and I was able to make an observation on the instrument there. I noted that the sun's image which came through the pinhole – there was no lens, of course – a hundred feet away, was of sufficient size (the sun being 30 minutes across) to enable you to make a measurement to about $\frac{1}{4}$ of a minute of arc. So you could get very accurate measurements with these instruments and you could make very accurate tables, which they did. The Arabs did not change the picture at all. They accepted the earth-centred picture as improved by Ptolemy, they accepted the eccentrics, they made certain observational improvements, but they did not revolutionise astronomy: they merely kept it, they polished it up.

Chemistry

This was not true of other fields of science, however. They did a great deal, for example, in chemistry, a region where they did not simply have to polish up what the Greeks had done, because the Greeks did not have any chemistry. You hear a great deal about Greek physics but nothing about Greek chemistry because there was none. The nearest you get to it is in the fourth book of *Meteorologia* of Aristotle, which is generally considered – though this does not really matter – to be a fake or spurious book. It is an old fake and it could not have been made much later than the second century AD. *Meteorologica* contains the idea of four qualities – hot and cold, wet and dry – that determine the four elements – fire, air, water, earth. Thus hot dry = fire, hot wet = air, cold wet = water, cold dry = earth. The book proceeds to describe the effect of heating and cooling, wetting and drying on various substances and thus gives a very rough picture of chemistry.

The real origin of chemistry, however, came from China. I will not say much about chemistry here, but I think I must at least explain its origins. It came from a concept the Chinese had. In the West, the Old Stone Age people used to use iron oxide, red ochre, for imitating blood; they rubbed it on the bones of their dead. But in China, more because of the lucky chance that there was a good deal of it there, they used mercury sulphide instead – vermilion, which is a much brighter red and better magic blood than you get from iron oxide. They also studied it and found that if you heated vermilion, two things came out of it, a yellow material which burnt and a shiny metallic material which ran. These are the true elements, they considered, the three primary things: the shiny thing which ran they called the female principle, the Yin: and the yellow material which burnt they called the male principle, the Yang; and if you combined the Yin and the Yang you get the combined principle which was the elixir of life, which was blood. Now this sounds very crude, but it is the basis of the whole of chemistry. If we now turn it into our physical terms, we are dealing with a superfluity of electrons in mercury, a lack of electrons in sulphur, and a balance of electrons – the sulphide. And, literally, chemistry grew out of what were called the three first things. This is a Chinese idea and not a Greek idea. It did not come to the West until the Dark Ages and seems to have crossed to there through Mohammedan channels.

Optics

The other field in which Mohammedan science made a great advance on that of the Greeks was that of optics, and this was very largely a by-product of medicine. Optics may be used in astronomy, as in the principle of the sun-dial and so forth but, in fact, it comes much more into the doctor's province, especially in the tropical and sub-tropical countries where there is an extreme prevalence of eye diseases of all kinds. This goes right back to the Babylonians, who were doing operations, cataract operations, between 2000 and 2500 BC. Therefore, they must have studied the eye and they introduced into science something very important which the Greeks did not have at all – the lens. The Greeks had the mirror and they knew that the mirror could focus the rays – that was due to Archimedes – but they did not have any lenses.

This was partly a technical fault, for to make a lens you must have transparent materials, you must be able to work with transparent materials in a big way. The commonest transparent material they could find, though still very rare, was rock crystal, what the Greeks called *beryllos* or *krystallos*, and they did make rather crude magnifying glasses with pictures drawn on the sides so that they seemed to be enlarged: you can see some of these in the British Museum. But it was only in Arab times that lenses started to be used. Then, the lens was considered to be a very remarkable thing indeed. The ordinary man had no idea what a lens could or could not do but the main concept was that it enabled you to see a long way. Friar Roger Bacon was one of the later people who was credited with being able to put up his lens in Oxford and see what was going on in Paris – here was a real television apparatus! This illustration (Fig. 28) from Roger Bacon's *Opus Majus* shows the curvatures of the refracting media of the eye. It closely follows that of his master, the man who really founded optics, Alhazen of Egypt. It is really a diagram, all made in the characteristic way of the times – everything had to be in a circle. There was no attempt to draw it as it was, but as it ought to be. There are two aspects of the eye showing a rather crude picture of the lens, the coats of the eye and the retina.

The Arabs found a most important thing which gave them the whole concept of optics: that you could magnify with the lens. What is more, an old man, no longer able to read, could read if he used one of these lenses. For a long time they used them as hand lenses and then some bright person – no one knows who – thought that you could put the lens in a frame and actually wear it, using both eyes and not only one, and thus, a little later, spectacles were developed. This was somewhere in the thirteenth century and it happened in China at the same time as in Europe. The early spectacles were not made of glass but of transparent precious stones, particularly beryl – hence the German word for spectacles, *Brillen*.

What really interested them on the optical side were the optical phenomena such as the rainbow. Roger Bacon introduced a new idea about science, the idea that you could study

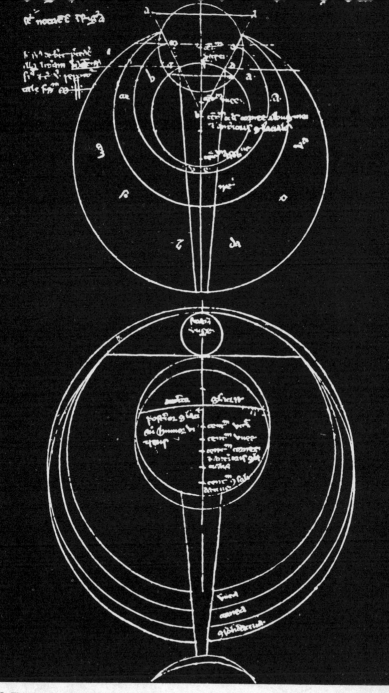

28 Thirteenth-century diagram showing the curvatures of the refracting media of the eye; from Roger Bacon, *Opus Majus*

by means of experiments. He wrote two very important books about it, a big book and a little book – *Opus Majus* and *Opus Minus* – showing how experiments could be used to explore the known world. This idea, in a sense, had been used by the Greeks but not consciously and consistently. It matters a great deal if, when you do various things and do them very well, you do not always notice what you are doing because you have no theory of what you are doing. For example, in my own field, X-ray crystallography, we used to work out the structure of minerals by various dodges which we never bothered to write down, we just used them. Then Linus Pauling came along to the laboratory, saw what we were doing and wrote out what we now call Pauling's Rules. We had all been using Pauling's Rules for about three or four years before Pauling told us what the rules were. Similarly, experiments had been done by the Greeks but they had never consciously adopted the idea of experiment. Now this, actually, is what Roger Bacon says about it. He is talking about optics: 'What I have now touched upon in general I wish to show in particular, by disclosing the basic principles of this very beautiful science. It is possible that some other science may be more useful, but no other science has so much sweetness and beauty of utility. Therefore it is the flower of the whole of philosophy and through it, and not without it, can other sciences be known.' Then, of course, he refers to Aristotle: 'We must note, moreover, that Aristotle first treated of this science. . . . After that, Alhazen treats the subject more fully in a book which is extant. Alkindi also has arranged some data more fully, likewise authors of books on Visions and Mirrors.' This referred to the use of optics for optical tricks like Pepper's ghost with which you could make things 'appear' that were really not there.

Of experimental science, Bacon says:

Since this experimental science is wholly unknown to the rank and file of students, I am therefore unable to convince people of its utility unless at the same time I disclose its excellence and proper signification. This science alone, therefore, knows how to test perfectly what can be done by nature, what by the effort of art, what by trickery, what by incantations, conjurations, invocations, deprecations, sacrifices that belong to magic, mean and dream of, and what is in them, so that all falsity may be removed and the truth alone of art and nature may be retained. This science alone teaches us how to view the bad acts of magicians, that they may be not ratified but shunned, just as logic considers sophistical reasoning.

It was not that he disbelieved magic, but he disapproved of magic.

This science has three leading characteristics with respect to other sciences. The first is that it investigates by experiment the notable conclusions of all those sciences. For the other sciences know how to discover their principles by experiments, but their conclusions are reached by reasoning drawn from the principles discovered. But if they should have a particular and complete experience of their own conclusions, they must have it with the aid of this noble science (experimental science). For it is true that mathematics has general experiments as regards its conclusions in its figures and calculations, which also are applied to all sciences and to this kind of experiment, because no science can be known without mathematics. But if we give our attention to particular and complete experiments and such as are attested wholly by the proper method, we must employ the principles of this science which is called experimental. I give as an example the rainbow and the phenomena connected with it, which nature are the circles around the sun or of a star, which is apparent to the eye in a straight line . . . and the circle is called a corona, phenomena which frequently have the colours of the rainbow. The natural philosopher discusses these phenomena and the writer on perspective has much to add pertaining to the mode of vision that is necessary in this case. But neither Aristotle nor Avicenna in their Natural Histories have given us a knowledge of phenomena of this kind. . . .

This, then, is really the beginning of looking into nature and finding out things by such looking – and by looking in special ways. Medieval scholars did, in fact, find out quite a lot about the rainbow and some of them, such as Robert Grosseteste, Bishop of Lincoln, maintained that, actually, they had the whole story.

Medieval technique

A characteristic of the fall of the Roman empire was the breakdown of the large engineering works. However, to make up for not having them, the medieval people had small ones that showed a greater ingenuity. The great achievement of

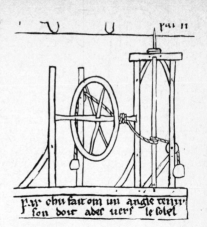

30 Villard de Honnecourt's
rope-escapement, c.1250

both Mohammedan and Christian architecture was to build in a very large space and cover it over using small stones or tiles. The Egyptians had used heavy stone beams as much as forty feet long, which they heaved up somehow with the help of many people. The Greek temple, again, was a pillar and transom architecture. The Romans built vast heavy arches and vaults. But the Mohammedans, following the Greeks, built their domes out of light brick and the northern Christians built vaults of great complexity out of small stones – the largest stone that you see in any cathedral vault is not more than two feet in any direction. As they improved their techniques, they made the stones smaller, and developed the system of fan vaulting. Fig. 29 shows, as it were, the finishing off of a cathedral. It was drawn by Villard de Honnecourt and is a working drawing. He had a pile of notebooks into which he put ideas. Fig. 30 is a very crude drawing of a very interesting machine which is called an 'apparatus by which an angel can turn his finger towards the sun', in other words, it is a kind of clock with an angel turning round. Well, we can forget about the angel – anyone can put in an angel – it is the machinery that is really interesting. The angel sits on the pivot at the top and the rest is an escapement; not a very complicated one. The descent of the weight and the uniform rotation of the vertical shaft is controlled by the to-and-fro oscillation of the fly wheel, around whose horizontal axle the rope is wound. It is a kind of crude clock escapement, all made of wood.

29 Drawing by Villard de Honnecourt of Gothic buttresses, 1235

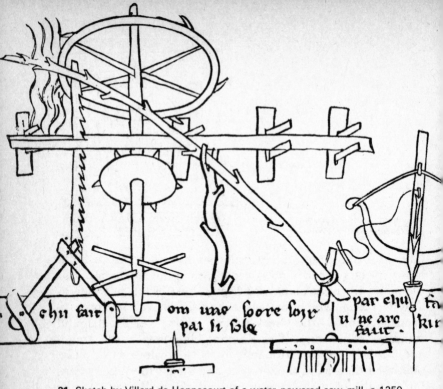

Within the image: chu fair · om uno boore forr pai li ble · par chu fa u ne arc fait · kir

31 Sketch by Villard de Honnecourt of a water-powered saw-mill, c.1250

Fig. 31, also from Villard de Honnecourt, shows a very interesting piece of apparatus, which seemed at the time to have a future. It has all the elements of modern automatic machinery in it. It is a sawmill. Now, a sawmill, right up to the last or even the present century, was made by digging a pit, building the beams above it, putting in a top sawyer and a bottom sawyer and just sawing. But even in the thirteenth century they thought that, really, this could be done without using human labour. Now, this machine, rather sketchily drawn in the illustration, is driven by a water wheel mounted on the main or driving axle. It carries at its lower end two spokes which are really acting as a kind of ratchet and cam arrangement. They press on a lever system that pulls the saw down. Then the saw is lifted up by means of a very familiar looking apparatus – it is just a bent tree! The tree does the up stroke and the mill does the down stroke. But the interesting thing is that the machine also advances the work. This is achieved by a wheel with spikes in it, mounted on the main axle which catches the wood and drives it on. These apparatuses were not very popular among sawyers and their unions and were mostly broken up. Thus power sawing did not really come in until our own time, although they had the idea then. And they had the idea, too, of a lot of other mechanical devices, some of which were absolutely decisive in changing the whole character of their civilisation.

Figs. 32 and 33 are from a slightly later period, the fourteenth century, and show two sorts of lathe, the pole lathe, again using the same principle with a bent, springy piece of wood providing the return stroke. Note that the work is turned between the centres and the operator is just holding the tool to turn: he has no slide rest. The other picture shows an even simpler type of lathe, it is a lathe for making beads for rosaries, a bow lathe. The monk is holding the bow in his right hand and is just about to spin it. This kind of mechanics was already well developed in classical times.

32 Foot-operated pole lathe; from the Mendelschen Zwolf Bruderbuch, c.1395
33 Bow lathe for making rosary beads; from the Mendelschen Zwolf Bruderbuch, c.1390

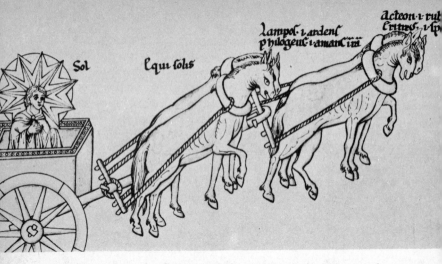

34 An early thirteenth-century drawing showing the horse harness; from Herrad von Landsperg, *Hortus deliciarum*

The new horse harness

Fig. 34, however, though it is a very curious subject for a picture, is really what gave the West its technical supremacy, although it did not really arise in the West but in China: it is the horse-collar. It is true that the horse first came into civilisation quite a long time earlier, about 2000 BC, and by that time the methods of harnessing oxen with yokes were already known. Yet the people who brought in the horse had no idea of harnessing it. They rode horses but they did not know how to harness them. Therefore, when they did they harnessed them as they had previously harnessed oxen, and it so happens that such a harness does not suit the horse very well. Consequently, Roman cavalry and Roman transport were incredibly inefficient. Can you imagine a horse without shoes, with a kind of strap across its neck, no traces, no collar. If you rode on it, you had only a saddle cloth, you had no stirrups. As a result, a Roman cavalryman could not charge at all because the moment he hit anything he was knocked backwards off his horse. Nor could he shoot a bow except forward: if he tried to turn round and shoot he would also fall off. One of the greatest inventions – which, again, came from China – was the stirrup, and the great success of the Huns and other people from the eastern plains was that they could shoot backwards as well as forwards because they could sit in a saddle on their horses. Later on, the enormous effect of the Norman cavalry was again due to the stirrups, by which they could hold the lance and actually push with it instead of having to throw it, which is all it had previously been possible to do.

Much more important, however, than all the purely military uses, was one of the elementary uses for a horse – for ploughing. Now, in the time of William the Conqueror, the standard tractor was an eight-ox plough, and the amount of a 'hide' was the amount of land you could plough with an eight-ox plough a day – and it was pretty expensive running an eight-ox plough. Later on it was found that you could do as much ploughing with four or even two horses. Horses took the place of oxen over much of the area for ploughing and also for drawing carts. Fig. 34 shows a complete arrangement with horse-collars and traces – this is a picture of the chariot of the sun – but the essential feature is the new horse harness. That new horse harness made it possible to plough uplands. Level land can be ploughed pretty well with oxen, but with horses quite considerable slopes can be ploughed as well, and the area under the plough almost doubled, as did the corn production in countries like England and France. And that invention, as I have said, was one of the first technical inventions from China which really transformed things in Western Europe. It was not of the same consequence over most of China and Asia because they did not have the same problem for irrigated fields.

Fig. 35 illustrates another key development. All through classical times, spinning was done with the free spindle. The idea of the high-speed drive of the spindle was one of those inventions which changed the whole situation, because it must

35 The invention of the high-speed drive for the spindle, c.1300

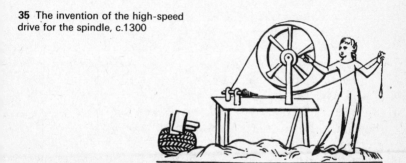

be realised that in all early civilisations the main item of trade was cloth of one form or another. If you could make cloth, if you had spun thread in the first place and set up looms in the second, you could easily transform the economy. Our great Industrial Revolution arose mainly out of the method of spinning, as you will see later on, because this medieval wheel with one spindle was turned into the huge machine with multiple spindles, although the basic idea was exactly the same.

Magnetism

I will now pass back again to another branch of science which appeared in this intermediate period and which proved to have enormous future development, that is, magnetism. The Greeks used the magnet – in fact they called it a magnet because it comes from the district of Magnesia in Asia Minor. They knew it attracted iron, and the attracted bits of iron would attract others, and so on. In effect, they knew the principle of induction, but what they did not know was that there was any relation between the magnet and direction. This is one of the greatest discoveries: in fact, I would say, without doubt, that it is the greatest discovery, in view of the human difficulty of making it, in the whole of the history of physics. Because, you see, it was one of these things that could not be predicted. No one was in a position to say that if you take a magnet and suspend it freely, it will point north and south. First of all, why should you suspend a piece of magnet freely? There was absolutely no reason to do so. Then, why should it point north and south? What is there north and south that concerns a magnet? There is no *a priori* connection between the two.

How the discovery came about is one of those little jokes of history, similar to what I was telling you about earlier in connection with the Greeks and their concern with polyhedra. The Chinese, too, were given to various types of what we call 'mancies', kinds of magic for finding out the future by means of observing the behaviour of things. One of the things they did was to spin objects round and see how they pointed when they came to rest. According to the direction in which things pointed, or according to the person at whom they pointed, this told them how to work things out. The Chinese were

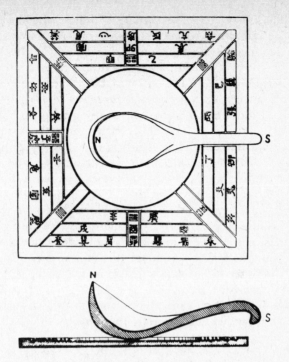

36 A reconstruction by Wang-Chen-To of the earliest form of compass. The balanced spoon was cut from lodestone (magnetite). The geomancer's board shown is of Han times, about AD 100.

always particularly careful about where they were buried because, after all, you only live a short time but you are in your grave a long time, so that the site of your grave is much more important than the site of your house. They had geomancers – at least, we call them that though the Chinese do not – who made a living out of going around and choosing the right place for people to have their graves, the fee charged being according to the wealth and importance of the person. One of the ways of determining where a grave should lie was to spin an object on a board. A geomancer's board (Fig. 36) had a circle in the middle and was divided into eight sectors each ornamented with eight arrangements of dots and dashes which was really a kind of Morse code, and all the combinations and permutations of the three short and the three long marks made these eight symbols the sacred directions. The essential thing is that on this board was put something to be spun.

In fact, they may have spun a Chinese spoon; you have probably all eaten with these spoons at Chinese restaurants. They balanced the spoon on its bowl and, as the handle is short, it spins very nicely. I am not prepared to say that all the ones you get in restaurants will spin, but you can get ones which spin very nicely. And we know from many texts that they used spoons in this way. The reason for using a spoon was purely astronomical. You know the constellation the Great Bear, which fixes the Pole. The Americans call it the Dipper, and we call it Charles's Wain or the Great Bear. It does not look very much like a great bear, but it does look like a dipper, and the spoon really represented that.

As the art improved, it was obvious that you would get better results if you made the spoon out of more precious materials, and you could make it out of jade or one of the other sacred stones, rock crystal, magnetite or lodestone. But, whereas all the other stones gave very variable results, they noticed that the lodestone did not vary – it always pointed south. And so the compass was born. The proof of this is found in the compass table, which is exactly the same, without alterations, as the old geomancer's table.

The compass was discovered in about the sixth century AD and it gradually, very slowly, got round to the West. We know roughly the time of it because there was an earlier method of navigating which did not use this kind of apparatus, but a rather more tricky one which we know from the Bible, from the story of Noah. But there are even earlier records; there are seals known from Harrapa in Pakistan dating from about 3000 BC, showing a ship over which a bird is flying. Now, how they came to this applied zoology, I do not know, but it is very, very old indeed. You carry birds on the ship and when you are far out to sea and out of sight of land, you release the birds. The birds, being cleverer than we are, immediately see the direction of land, fly off in that direction and you follow. This is how a Viking named Floki found Iceland in about AD 900. In one of the manuscript versions of the story (which was not written down until some three hundred years later) the writer inserted a little explanatory note. He said that in those days they used ravens because the use of the lodestone was not known. That gives us, as it were, an upper date: it must have been very well-known by about 1200 and certainly not known

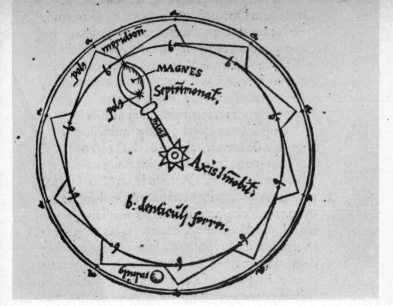

37 The use of a magnet to make a perpetual motion machine; from Pierre de Mericourt, *De Magnete*, 1269

by 950, so the true date must lie somewhere between the two. The discovery of the compass proved to be the clue to the other great discoveries that I will talk about in the next chapter: the crossing of the oceans, the discovery of longitudes and the whole of modern navigational science, and the revolution which that navigational science brought to astronomy and physics.

About 1260, there was a remarkable man called Peter the Pilgrim (de Mericourt), and he wrote the first scientific treatise of the time, on the magnet. He did not know that the magnet points north and south, but he did know many other things about the magnet: that it had poles; that if you cut a magnet in two you get two magnets each with two poles, and so forth. Immediately, of course, people thought of ways to use it. Fig. 37 shows one way to use the magnet to make a perpetual motion machine. There is a kind of armature of iron around the outside and a piece of lodestone held in a grip inside. The lodestone attracts the piece of iron on the armature and makes the whole wheel go round in perpetual motion. The reason no one ever bothered to do it to prove it would not work, was probably because they had other excuses for saying why it would not work.

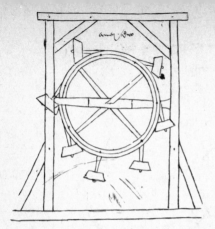

38 Perpetual motion: the use of weights to make the machine go round; from the notebooks of Villard de Honnecourt, 1235

Fig. 38 shows a more elementary type of perpetual motion of the kind that inventors are still making, where the weight flops over and one side of the wheel is always heavier so that it has to keep going round – this, again, is from Villard de Honnecourt's notebooks. People had the idea of generating power, but they still thought it could be done by some kind of trickery.

Fig. 39 is a little on the theoretical side. To measure altitude they already had quadrants with degrees on them, but for more practical use they invented in the eleventh century this rather simple device, the cross-staff; the illustration is from the text-book *How to find the height of a tower from a distance.* You move the staff backwards and forwards until it just covers the angle you want to measure: you get two observations, you see, from the cross-staff – it is graduated in this way. This was all they had on ships in the twelfth century and it was considered to be a very modern and most unsailorlike instrument. The sailors made fun of it, saying to the navigator 'What are you trying to do? Are you trying to shoot the sun?' Taking an observation right up to the days of radio was called 'shooting the sun' because the cross-staff looks so very much like a crossbow.

Fig. 40 shows the chief instrument of the outdoor astronomer. It is a rather interesting one and is actually in a manuscript of Chaucer, who was very interested in this kind of thing. It is an astrolabe, really a calculating machine which enabled you, by making one or two observations, to find the time of the night if you knew where you were, or, to find where you were if you

39 How to find the height of a tower from a distance

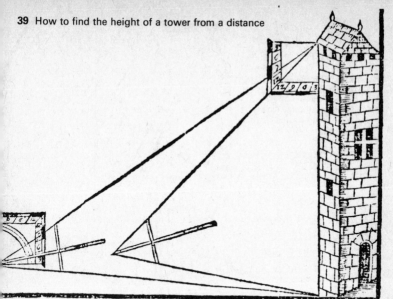

Quand vous voudrez mefurer felon la couftume vulgaire, la hauteur d'au
ne chofe, vous mettrez le petit Bafton fur la premiere ou feconde entre
coupure

40 (below) Chaucer's astrolabe

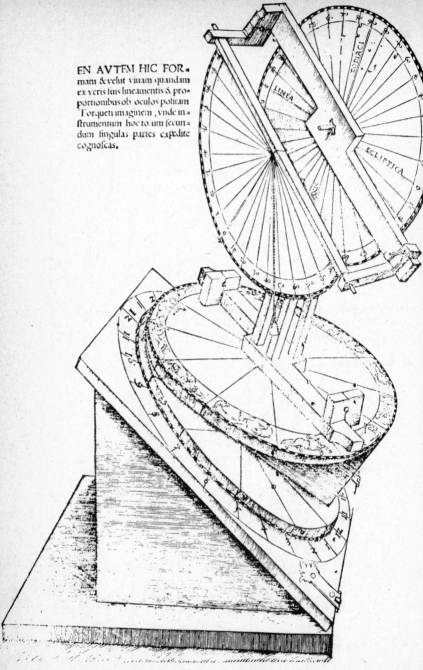

EN AVTEM HIC FOR-
mam & velut viuam quandam
ex veris suis lineamentis & pro-
portionibus ob oculos positam
Torqueti imaginem, vnde in-
strumentum hoc totum secun-
dum singulas partes expedite
cognoscas.

41 The use of graduated circles in astronomical observation

knew what the time was. The picture shows one side of the astrolabe. The astrolabe was the general instrument: you could have an astrolabe up to about a couple of feet across but not much bigger, so the accuracy was very low indeed.

Fig. 41 shows a more sophisticated instrument of the late Middle Ages, which is an attempt to combine a whole number of graduated circles together. On this apparatus quite a number of astronomical observations could be made much more simply than with an astrolabe, which would only enable you to take a simple altitude one way and an azimuth the other.

This chapter has given a very cursory picture of the intermediate medieval times, and I will sum it up by saying that they saved a great deal of the Greek contributions in physics. In other fields, sciences of great importance were reformed, for instance, chemistry and to a large extent medicine, but to physics the chief thing they added was information on optics and magnetism. In the next chapter I shall come on to the revolution which followed this and laid the foundations of what we call modern science.

6

Heliocentric Physics

The next section in this history is, in a sense, the most important of all and the one which connects up with physics today. It is very necessary to understand the stretch and extent of the phenomena that we now call the First Scientific Revolution – you may not know it, but we are now living in the Second Scientific Revolution. We call it the First Scientific Revolution because, despite the Greeks' remarkable mathematical feats, what we call modern science did not really arise in Greek times. The First Scientific Revolution is associated with such tremendous events as the discovery of the nature of the solar system, the discovery of universal gravitation, of the properties of light, of the vacuum, of gases: in fact, with discoveries involving all the main branches of physics as we know them today. But it is also associated, in a very different way, with the events of the time. It is a peculiarly European movement though, of course, it has now spread all over the world. One of the great historical problems, which has not been solved and may not yet be susceptible to solution, is to find out why this Revolution happened when and where it did. We can offer pretty good negative reasons: we can see very good reasons why it could not happen in different circumstances but that is not sufficient reason. It is a necessary one but not sufficient.

We know that in the absence of certain factors – interest, money, technical abilities, for instance in mathematics – the First Scientific Revolution could not have taken place. But we do not know why it did not happen in classical times which were as wealthy in different parts of the world, such as India and China, as in Europe. These areas had an abundance of extremely clever men, men of genius, and yet they did not take this particular step. It is associated in Europe with two other phenomena of which you read in the history books, and a third one which you may not necessarily find there. These are the Renaissance and the Reformation, producing the event which fits the transformation better than the Reformation, namely, the rise of capitalism.

One of the largest contributions to the Scientific Revolution was what might be called the artistic revolution, which we usually call the Renaissance. When one thinks of the Renaissance, one thinks of it essentially as an artistic revolution, a revolution in painting, in architecture, in manners of life, in poetry, in courtesy and so forth. Together these represent a change in the economy, in technique, in attitudes towards the world which were reflected in changes in religion as well as in the physical appreciation of the world.

Now, what do all these mean taken together? I cannot give here a historical analysis; it would be wrong to do so, but I need to sketch in enough background to see where the different parts fit in. The Scientific Revolution, essentially, moved in stages. We can discern roughly three major ones: a preparatory stage which – by convention more than anything else – starts in 1453, the date of the fall of Constantinople, the end, as it were, of the last continuity of Greek culture. This preparatory stage goes on until, let us say, 1543 – another useful date because it is the year of Copernicus's revolution which destroyed the old conception of the world, of the central place of the earth in the universe. It is essentially a critical stage, one where the old system was examined and found faulty and where the lines of the new system were worked out.

The next stage which is, as it were, the crest of the wave, is the Renaissance, followed by the Reformation carrying us on to about 1620. That period covers the filling out of the picture of the solar system by Tycho Brahe and Kepler, and it also covers the development of the telescope and the optical and dynamical discoveries of Galileo. The central period of the seventeenth century takes us to 1687, the date of Newton's *Principia*. With its publication, all seemed to be tied up and codified. It was such a decisive achievement, in fact, that it had the effect of marking a definite break in development. Once Newton had completed his work, it really did not seem at the start that it could be much improved on; so people did not try to do so for another hundred years. And all this built up to the First Scientific Revolution. I forbear to mention here other aspects of it, for example, the biological aspects, although the period we are considering includes Harvey's work on the circulation of the blood.

The fundamental advances were discoveries that destroyed the Greek world view, first, through the development of anatomy, the working of the human body, and second, through the discovery by experiment and observation of how the solar system works. Previously, things above, celestial things, were for observation only and the people on earth could not be expected to understand how such divine objects moved. This was expressed by saying that they were moved by angels or they were moved by celestial mechanisms, at any rate they were moved by some kind of mechanism which we did not have on earth.

With the developments of this period, particularly the work of Kepler, we get the picture, which was completed by Newton, of the dynamics of the earth being transferred to the heavens. Incidentally, this work completes the circle to the extent that the discovery of how the heavens work was considered as the main object of science, physics. This was because the motions of the heavens were considered to be intrinsically at the beginning of the problem, the most important things there were in the universe. The whole fate of life and everything else was tied up with the movement of the heavens: the heavens ruled the earth. Therefore, whoever understood how the heavens worked, would understand everything on earth.

By the time Newton's work was understood, it was realised that the heavens do not rule the earth, that the problem was, in fact, not a terribly important problem and its solution likewise not important. What was important was the means of arriving at the solution. In arriving at the solution Newton discovered – or Galileo and Newton together discovered – the new science. (Galileo called it the two new sciences, one being dynamics and the other statics.) They also discovered the corresponding mathematics, the mathematics of the differential equation, which has been the basis of the whole of theoretical physics ever since. This was not the original object of the exercise, however, it was simply the means of solving the problem of the structure of the universe but then, of course, it found other uses.

In this second period, comes the beginning of the foretaste, as it were, of the Industrial Revolution, the steam engine, etc., in the study of the vacuum. In Table 1 (p. 37) are set out the discoveries of Copernicus, Galileo, Kepler, Newton. But there is someone else who should be put here as he was part of the general picture and, in a real sense, its originator; that is Columbus in 1492.

The change in Europe

We will now go back a bit and think about Europe in this period. I say 'think about Europe' because the change occurred in Europe and at that time did not occur anywhere else, for entirely historical reasons which we still need to explore. The Mohammedan world had taken a terrible knocking from the Tartars. It was much nearer to their base, Mongolia, than was the Christian world. By the time they had reached the Christian world, the Tartars had run out of steam, so to speak, or at any rate they had reached as far as they wanted to go by about 1280. They had just penetrated the boundaries of the German empire, no further; but they had destroyed and massacred the people of Baghdad and they had generally 'chewed up' the Eastern world. This included a great part of India, China and Central Asia. The Eastern world – except, curiously enough, one country, Egypt, which was just out of reach – had been so heavily battered that it was not likely to be the source of any great scientific advancement at that time. Although the Tartars gave way to the Turks, the Turks did not develop much of whatever they inherited, because there is more in recovery than merely recovering from battles. There had been a whole attitude towards life and this had been shattered long before by the first rising of the Ottoman Turks in the Near East in the twelfth century. The idea of progress, the idea of finding out new things, was being actively repressed then in the interests of orthodox religion.

In the meantime a great deal had happened in the West. Now, the West was a backward region – what might be called an under-developed region – although it was already building a civilisation of its own. The most under-developed part of that region happened to be Britain because it had never had much civilisation. It had been on the edge of the Roman empire and, consequently, benefited less from Roman civilisation than anywhere else. But the important thing about Britain, and to a lesser extent the neighbouring countries of the Lowlands – Holland and Flanders – was that they were on the edge of the continent, bordering the sea.

To a very large degree, the whole process about which I am about to talk is one that comes from the exploitation of the sea for two purposes, one is for trade and the other for food, for fish. One great revolution was brought about by something very small, the change in salinity of the Baltic Sea which drove the herrings out of the Baltic into the North Sea and made fortunes for the people in the Low Countries and Britain. This gave enormous encouragement to navigation. You may remember my saying in the last chapter that navigation was the sphere in which science could be most immediately useful, particularly navigation in the open ocean and not in the closed seas such as the Baltic and the North Sea as well as the Mediterranean. So navigation was, as it were, the key – or the key-hole – to the utility of science.

Meanwhile, the characteristic kind of economy that existed in Northern Europe was first of all a subsistence economy. The manorial system was composed of farms which produced nearly everything they needed except iron and salt. But in the thirteenth and fourteenth centuries they were short of people, of men, partly due to the Black Death. They had more land than they could cultivate, they had more things to do than they could find people to do them. This was the opposite of the state of affairs that had existed for a long time in the classical era. As a result, there was a premium on doing things mechanically. I have already mentioned that according to the Domesday Book there were some 5,900 water mills in England, and these mills not only ground flour but they helped to finish cloth by fulling.

Then there was another element – trade. In the first stages, there was no trade in the roughest of raw materials because they could not be carted about profitably. There was little trade initially in corn but there was a very considerable trade in cloth, and the most elaborate trade at that. The cloth trade, which started in Italy, gradually spread over the Low Countries and then to Britain. We had no one in Britain who could do any fine weaving until about 1350: we had to import weavers from Flanders to do the weaving and the fulling and other treatment of the cloth. England produced rough cloth. The fine cloth, as well as the glory of the late Middle Ages, came from the birthplace of modern science, Florence. Florence was, so to speak, the Leeds of Italy, the place where wool was worked up. The great Florentine houses, the Medici and so forth, all grew up in the wool trade. They developed fine wools; they obtained their rough wool from England, worked it up, dyed and finished it and sold it back to England. Just as later we worked up the Indian cotton and sold it back to India. The point here is that it was in those countries operating the textile trade where there was accumulation of capital sufficient to get the whole process of the Industrial Revolution going.

The Agricultural Revolution

Another feature of the time, which is not usually taken much into account and which we are only just beginning to understand, was what we call the Agricultural Revolution. It was really the biggest movement of its time and a movement can be very big without anyone seeing it. The Agricultural Revolution was essentially a way of keeping cattle alive in the winter and using the manure to improve crops. Normally, in the northern countries, they slaughtered as many cattle as were needed to carry them through the winter, salted as much as there was salt for, and then had a nice, religious fast all through the spring when there was little meat to be had anyhow. The development of this Agricultural Revolution came from the East, irrigation from Syria, and crops like lucerne, for fodder, from Italy. The Agricultural Revolution in Europe started in Lombardy and by say 1470, the national income of Lombardy – or the State of Milan – was equal to that of England and France put together. The wealth came from a combination of a steady supply of food, a much greater number of people and the success not only of the wool trade but of the metal trade, because the working up of metals had started again in a big way.

With regard to the metal itself, you cannot, of course, pick up metal anywhere, and one of the weaknesses of Italy, which prevented it from becoming absolutely the dominating power of Europe, was that it had no metal mines. The metal mines were in the Empire, in Austria and Styria mostly, and in Spain. Spain and Germany became the dominating powers in these

42 Beating out sheets of brass: from Biringuccio, *De la Pirotechnia*, Venice, 1540

first stages because they had metal; later it was to be Sweden and Russia. Metal-working developed a great deal of mechanical ability and this was a fundamental point which was recognised at the time. In the year 1556, a book written by

43 Mechanical wire-drawing. The water wheel drives a crank which on every half-turn enables the wire to be drawn through the draw plate. From Biringuccio, *De la Pirotechnia*, 1540

George Agricola was published. It was the first of a kind of scientific-technical hand-book on mining which included sections not only on making mines, getting the ore out and the much more difficult job of getting the water out of the mines, but also on the working up of the metal into useful things. It was a really comprehensive treatise and influences the much further development of the mechanical devices. Fig. 42 does not look very sophisticated at first glance but it is what is called a 'battering' and shows how to make sheets of brass. None were made in England until Queen Elizabeth licensed the first battery company in, I think, 1585, but they were very well known in Nuremburg. There they made brass plate and with that scientific instruments and, afterwards, maps. Nuremburg became known as the geographical centre of the world simply because it could make better brass plate. Iron plate was mainly needed for armour.

Fig. 43 is very interesting. Here you really see power at work; it is a wire-drawing mill, a rather simple, one-man enterprise. The waterwheel is there, the power; the man on a

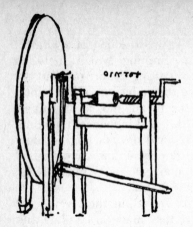

44 Fly wheel, crank operated lathe; from the *Codice Atlantico*, Leonardo da Vinci

very convenient sling with his self-acting clips. He simply waits until the crank moves forward, advances the clip and lets it grip; when the crank moves back the wire is pulled through the draw plate. This is a very old way of doing it but it still persists. There are several parts of Africa where they will not accept wire unless it has the jaw marks on it because it is not genuine to them without this. Wire is a very important auxiliary, especially needed for wool carding. I draw particular attention to two things on Fig. 43: there is not only a water-wheel but a crank and, as you can see, the crank is a piece of forging and, clearly, you have to have the mastery of iron working before you can make such a machine.

Fig. 44 shows the crank again and this, incidentally, is by a great master. It is a drawing made by Leonardo da Vinci himself of a fly-wheel, crank-operated lathe. It is still a foot-pedal lathe and I remember that I did some of my own early

45 Articulated driving chains; from the *Codice Atlantico*, Leonardo da Vinci

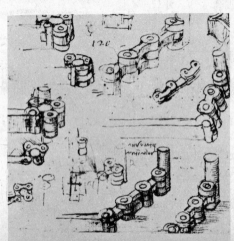

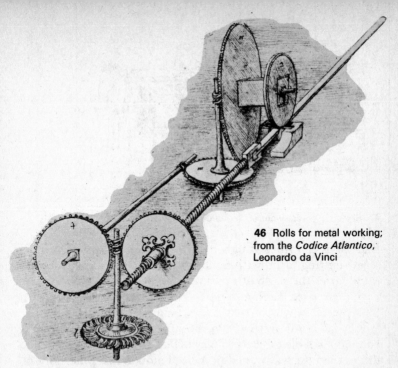

46 Rolls for metal working; from the *Codice Atlantico,* Leonardo da Vinci

scientific work on such a lathe that was actually not so good as da Vinci's because it had no fly-wheel.

We have now reached the beginning of the mechanical age and Leonardo da Vinci has left us many drawings of that period. Fig. 45 shows something very familiar, a chain gear, but it is not very original. As Needham has shown*, the Chinese had made a beautiful clock about 400 years before Leonardo da Vinci in which the main drive was by a chain made exactly as depicted in this illustration. It does not follow that everything that was drawn by Leonardo was invented by him. He drew what he saw besides what he himself conceived.

Fig. 46 shows some rather sophisticated gearing, water-driven by turbine. You will notice a vertical-axle turbine and not the conventional horizontal-axle waterwheel. The machine has two elements in it: the rolling mill element, the roller; and the advancing element, geared down one step further and working through not only a wheel but a long screw-drive. To get all this to work would require quite accurate machining.

*J. Needham *et al., Heavenly Clockwork: the Great Astronomical Clocks of Medieval China,* Cambridge, 1960.

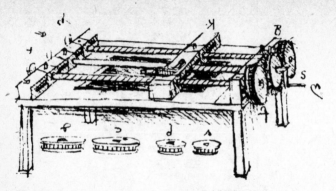

47 Sketch of a screw cutter: Leonardo da Vinci

Fig. 47 is a rather rough drawing but it shows a complete screw-cutting lathe with a variety of gears. Varying the gear-ratio varies the pitch of the thread. The cutter is marked with a Я, the two leading screws are on the outside, and the workpiece, on which the screw is being cut, on the inside.

Fig. 48 is interesting, too, because it is a bit of nonsense; you may be able to see through it. They were very keen on such things then – a really good perpetual motion machine. This one is for grinding knives and armour and similar things. There is a water-wheel which works the main-drive and at the same time operates an Archimedean screw which lifts up the water; the water then runs down and operates the wheel. It is beautifully drawn, but no one has actually been able to make it work! Yet, it shows a degree of gearing, all of which comes from mill work. This type of gearing is not new, it was old in Greek times, but it was developed very much further by a whole set of people who worked at these things, the so-called mill-wrights. They wandered around the country not only making the mill wheels, but also arranging all the sluices, the dams and the lay-out of the waterways.

Fig. 49 is a very interesting example of a rolling-mill, semi-mechanised, with the horse-drive on a vertical axle. The working-piece is heavily loaded, the actual sheet is underneath the rollers. The gear-box, operated by the man on the left, is above and simply switches the main semi-vertical shaft over from one set of gears to the other. That kind of reversing mechanism was well adapted for relatively simple things and it is really a quite powerful, if only one-horsepower, machine.

Ein andere Werbel oder Schraubkunst mit doppelem ahngriff

48 A perpetual motion machine for grinding knives. The water wheel operates an Archimedean screw which raises the water which in turn operates the water wheel. From Georg Andreas Böckler, *Theatrum Machinarum Novum*, Nuremburg, 1673

49 Semi-mechanised rolling mill; from Zonca, *Nuova Teatro di Machine et Edificii*, Padua, 1607

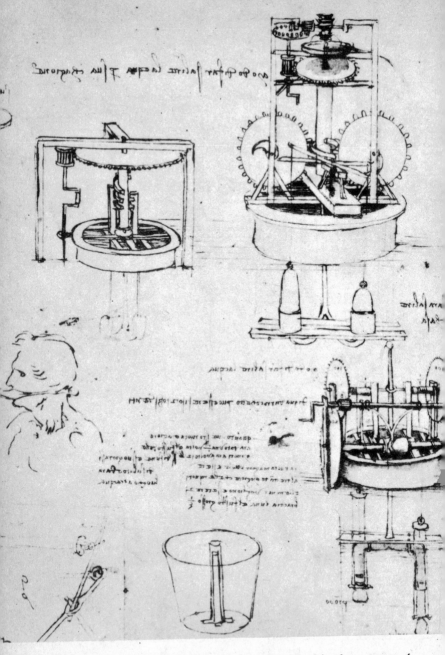

50 Sketches by Leonardo da Vinci mostly of double-acting force-pumps of various ingenious designs

In Fig. 50 we come to another type of mechanism which was very much used in those days, the water-lifting machine. The Greeks and Romans had quite sophisticated pumps but they were not anything like as sophisticated as those illustrated. These are sketches by Leonardo, mostly of double-acting force-pumps. Most of this was done for fun and games, the main object – especially in the Mediterranean area – being to produce elegant fountains and waterworks of various kinds for princes' gardens. Cost, therefore, was no object. But, as you will see later, these machines could be used for other things. Incidentally, note in Fig. 50 the very ingenious divided screw mechanisms and also the crank. These were not however very serious mechanisms; they were hand-operated.

Fig. 51, on the other hand, is rather attractive to amateurs of today. It demonstrates how to walk and breathe under water. It has not quite got the snorkel element but the rather simpler form which I used as a schoolboy. In fact, it was a rather silly thing to use because, of course, I did not allow for the other boys who delighted in pushing the cork under the water when I was underneath. But this idea of diving also fitted in with the great concern at the time for shipping.

51 How to walk and breathe under water; from G. A. Borelli, *De motu animalium*

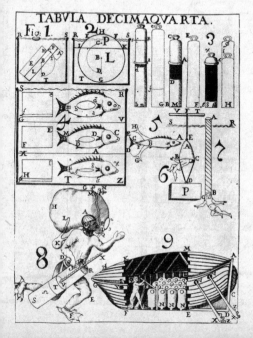

53 Chain pump and treadmill; from Agricola, *De re metallica*, Basle, 1556

Fig. 52 is more practical: it is from the great Agricola's book. It shows a reversing gear hoist operating as a way of getting water out of mines. After all, there is usually about ten times more water than mineral to get out of a mine and this illustrates a very simple way of doing so. The water is brought up in a large hoist operated by direct action from the over-shot water-wheel which is divided into two and it can be made to go one way or the other according to the operations of the crane man who is up in his hut. Another man would give him clues as to which lever should be heaved up. There was a hoist of that kind for the minerals too. So there is little new in what we see nowadays; it was all very familiar in Agricola's time. Incidentally, the mining areas in Germany were very suitable for this kind of apparatus because the ground was rough and there was plenty of water and wood. One of the difficulties for the people of classical times was the presence of areas which were dry through lack of permanent running streams. The Romans made the most wonderful mines in Spain with elaborately constructed turbines, with the most elegant curves. They knew all about the techniques of construction but they still had terrible difficulties with it. Try draining a mine by means of twenty Archimedean screws each of which had to be geared to a turbine situated elsewhere, and you would find it harder than the way it is done now.

Fig. 53 shows a rather rougher way of doing the same operation. Here we have the treadmill and the miners working it. Miners by that time – and this is a very important point – were not slaves any more, they were free men and they had to be given comparatively high wages because there was nothing to prevent them from setting up a mine of their own. In Cornwall, anybody certified as a 'free miner' could throw his pick anywhere he liked and had the right to dig for tin there and to keep the tin – apart from the share the king took out of it.

I have given you, as it were, a glimpse of the mechanical and industrial background of that period, but now let us look at the more scientific part of it. Nothing that I have shown you here is in principle beyond what the Ancients had. The new developments did not as yet involve fundamental discoveries. There was nothing new in the pump or in the gearing; the essential thing was that more people knew how to use them. They had gearing for small objects too, for clocks. Their clocks were worked on a rather simple system with what is called a folliot which gave, more or less, synchronisation. They were good enough because the medieval clocks were not really expected to tell any accurate time, no one bothered about any period less than an hour. When they wanted to, they could always use a sand hour-glass.

The revolution in astronomy

Where then did the new ideas come in? In the first place, they came in with the revolution in the understanding of the heavens. I mentioned earlier the Greek picture of the heavens, which persisted even in the greatest detail right through the Renaissance. The work of Ptolemy, which summed up that of many other Greek astronomers, was a very precise analysis of the movements of the planets all conceived of as moving round the earth. I referred earlier (p. 96) to one of the devices they used to explain these movements, the eccentric. The second device was simply to put a rotating circle, the epicycle – which in reality represented the motion of the earth – on to the circular orbit (Fig. 54). This device was in fact geometrically equivalent to that of the eccentric. It struck a few astronomers as odd that all the epicycles were carried out once a year, but it did not matter whether it was odd or not – the essential thing was that it worked. As I have said, the Arabs made tables, and the Arabs conquered Spain, and most of their astronomy was done in the later periods in Spain, which was free from Turks. Then, the Christians – Crusaders and others – drove the Arabs out of Spain but they took over the astronomy and did not destroy all that had been achieved.

The first astronomical tables, the so-called Alfonsine Tables, were made in Toledo on the basis of Arabic tables which were based on Syrian tables which, in turn, were based on Greek tables; but the observations went on. They had to do this because if the tables were not kept up with observations they would get further and further out, and this eventually was precisely what happened. The tables went very wrong but a certain improvement, curiously enough, was effected by one of the later Tartars, Il Kahn Ulug Beg, who had an observatory in Armenia and who made another set of tables, the Il Khanic Tables. These tables, too, soon became very insufficient.

Now, what were the tables required for? It was a very active time politically, with a new kind of ruler, a new kind of prince who did not entirely depend on muscle, so to speak, but also carried out very subtle diplomacy. He was very wealthy, collected taxes, and was very cultured. Naturally, he required a certain amount of science to conduct his affairs and the best science, the highest science, was astrology, and astrological computations, alas, usually went wrong. The observations, of course, were not accurate enough. First of all, the stars had not been sufficiently carefully observed and, secondly, the calculations had not been done well enough. A mistake might have been made in connection with a battle and one they counted on winning might be lost and this could be blamed on not really having spent enough on scientific research.

This state of affairs continued until 1610. Kepler was the

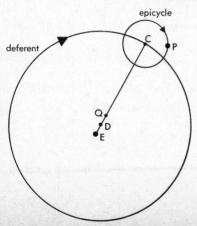

54 The device of the epicycle for the motion of a planet, P. The centre of the deferent is D which does not coincide with E, the Earth. The centre of the epicycle C does not rotate uniformly about D, but moves so that the line CQ moves through equal angles in equal times

mathematicus of the Emperor Rudolph II of Germany and his main function was to produce good horoscopes. That great champion of Christendom, Wallenstein, who fought the Thirty Years War which ended in 1648, had a horoscope drawn up for him by Kepler. Kepler did not believe in horoscopes overmuch himself, but he did point to what might be called the law of economic determinism when he said, 'God provides for every animal its means of sustenance – for astronomers he has provided astrology'. Nowadays, if you want money, all you have to do is to get into the nuclear field, or into making rockets, and you never have any difficulty in raising money, but in those days you could raise money from a much more harmless thing, from a horoscope. So the work on the tables went on.

Another point which is relevant here is that Ptolemy had written accounts of his work. They had been translated, as I have said, into Syriac, into Arabic, and from Arabic into Latin and, as a result, as with any translated scientific treatise, they were full of mistakes. They were also full of words which the translator, who was probably not very highly paid or very competent, very reasonably could not understand at all and he guessed what the words meant or left them in a totally incomprehensible form of Arabic. Now, when the Turks took Constantinople, the result was the flight of a large number of Greek refugees, rather like the German refugees in Hitler's time. They were widely dispersed and, as they were well educated and very cultured, they obtained work without much difficulty. For instance, they even obtained posts teaching Greek at Oxford, although Greek was very much disliked at Oxford – or, let us say, disapproved of. Students would read Greek testaments under their beds so that their tutors could not find them at it – it was considered a subversive language. Nevertheless, it was so beautiful and it so fitted the spirit of the Renaissance, that it prevailed. Thus they could read – much more important from our point of view – the original Greek texts, or fairly accurate copies of the texts of Archimedes, of Ptolemy, of Euclid, and that really made possible a fresh start in astronomy – not to do anything different but at any rate, to go over what had not been done properly.

There were a number of astronomers, firstly in Germany, who made observations and corrected tables. I do not need to burden you with their names – the most famous was called Regiomontanus – and astronomy became a very respectable and elegant occupation. Copernicus, apart from settling the nature of the solar system, occupied himself in translating a slightly improper minor Latin poet, the point being that the Renaissance gentleman was expected to do all these things.

Navigation

But there was another side to this activity, the other use of astronomy was in navigation. Here, there was a research school set up, around 1450, by Prince Henry the Navigator, who decided to collect all the books on navigation, astronomy and geography, and then see how to get where he wanted to go, which was where the money was. This was on the other side of the world, part of what we now call Indonesia – the Spice Islands. There was a great need for spices and the question was, how did you get to where they were to be found? If you tried to go the direct way it would have meant paying such heavy portage dues and other taxes, that it left very little profit in the end. The trade went through all the time but it was not a very profitable trade except, of course, for the Egyptians and other people on the route.

So the idea of gradually working a way round Africa, was put forward as the deliberate policy of a very impoverished, miserable part of Europe, Portugal. During this time things happened. The navigation started first of all as a kind of coasting. Actually, the way to get round Africa, if you want to do it quickly, is not to coast – or to coast only as far as Cape Verde – but to strike across, hit Brazil and come back on the other trade winds, landing in South Africa. However, this required sailing across the open ocean. For a long time no one believed that Africa could be circumnavigated. The Portuguese managed to do so by coasting and when they rounded the Cape in 1487, they found Arab navigators who pirated them all the rest of the way to India. The Arabs were just as advanced in navigation as the Europeans – and not only the Arabs but also the Chinese. There was a Chinese Admiral, Chêng Ho who, if he had not been recalled by the Emperor, would probably have established the Chinese empire in Britain because, slightly before Columbus he had a large fleet of about 4,000 men (Columbus, I think, had about 90 men). Chêng Ho's fleet sailed all the way round India, occupied Ceylon and went on to the coast of Africa. There they collected a giraffe and brought it back to Peking where the Emperor said 'I disapprove of this animal, it ought not to exist'. But maybe he disapproved of these journeys, thinking 'if an admiral is as powerful as that and can conquer a kingdom, how do we know that he will not set up on his own!' So Chêng Ho was recalled and an edict was given out in China that no further foreign sea journeys should be undertaken. So, as I say, navigation was not entirely a European skill – but it soon became so.

Columbus's connection with navigation was quite secondary, he was really a theoretical physicist. All the learned people knew that the earth was round, although they had varying opinions as to what was its diameter. As Columbus was trying to sell his project, he took the lowest possible diameter that anyone suggested because if it had been known how big the earth was no one would have financed his expedition. No one knew, or could know, that America lay in between. Columbus had to use all kinds of tricks to get the money. He cadged around for many years and finally raised some money for a small expedition from the Queen of Spain, Isabella, if for no other reason than to spite the Portuguese. The Portuguese had a monopoly of trade with the East and the Spanish hoped that this crazy man might possibly break the Portuguese monopoly by sailing round the world and getting to China by sailing westward instead of the other way round.

So Columbus set out on his voyage and although, as I say, he was a theoretical physicist, he did understand navigation – but he cheated as much as possible. What I mean is that he kept a very careful log but wrote down figures that did not correspond to what he had observed, so as to make his crew feel that they had not travelled as far as they had. This was because all the crew thought that it was an absolutely fatal expedition and that they would never get back again. If they had known how far they were going to go, they would have mutinied long before. They were, in fact, just on the point of mutiny when they sighted land. This landfall, of course, made all the difference, not only to history but to science.

Columbus also made some very interesting scientific observations as he crossed the Atlantic. He noticed that the declination of the compass (i.e. the deviation of the needle from the true north) had quite altered and this gave great hopes that they would be able to find their longitude from the declination. But even when they found that the declination varied from place to place and that it was really not a very effective way of finding the longitude, they still persevered with it. Columbus's key achievement was to open up, as it were, the idea of the round world. I will just give you one small quotation from a man of the time called Jean Fernel who is mostly known as a doctor – he gave the names to the two sciences that we know now as physiology and pathology – but who was also very interested in navigation and about 1540 measured an arc of the meridian in France. In 1530 he wrote:

But what if our elders, and those who preceded them, had followed simply the same path as did those before them? . . . Nay, on the contrary, it seems good for philosophers to move to fresh ways and systems; good for them to allow neither the voice of the detractor, nor the weight of ancient culture, nor the fullness of authority, to deter those who would declare their own views. In that way each age produces its own crop of new authors and new arts. This age of ours sees art and science gloriously re-risen, after twelve centuries of swoon. Art and science now equal their ancient splendour, or surpass it. This age need not, in any respect, despise itself, and sigh for the knowledge of the Ancients. . . . Our age today is doing things of which antiquity did not dream. . . . Ocean has been crossed by the prowess of our navigators, and new islands found. The far recesses of India lie revealed. The continent of the West, the so-called New World, unknown to our forefathers, has in great part become known. In all this, and in what pertains to astronomy, Plato, Aristotle and the old philosophers made progress, and Ptolemy added a great deal more. Yet, were one of them to return today, he would find geography changed beyond recognition. A new globe has been given to us by the navigators of our time.*

It is all very well when you find people saying today, 'we recognise the Renaissance, we recognise the Scientific Revolution' as events long past. But the people of that time also recognised them perfectly well. You could not have a more explicit sense of this opening out of the world and particularly this link between navigation and astronomy than is shown in this quotation.

*Jean Fernel, *Dialogue*, 1530.

The next stage was that reached by Copernicus. Copernicus was a very strange character. He was born in Poland, which had just partially thrown off the hold of the Teutonic knights. His uncle was a bishop who accorded him a very long and thorough education in Italy. He studied law and medicine, philosophy and *belles lettres*, and incidentally picked up a certain amount of astronomy. He was very interested, as people rarely were at that time, in navigation. But he was one of these very dry-water sailors for though he lived most of his life within twenty miles of the sea, he never went down to it. Nevertheless, he wrote a great deal about navigation. He looked into the whole system of Ptolemy, of which many criticisms were to be found, as were criticisms of Columbus too. Of Columbus it was pointed out that he should not have done it, that he did not really know enough, that he was a very bad navigator, that his ideas were quite wrong, that he just happened to have the good luck to get where he did. The same is largely true of Copernicus's ideas on navigation. Nor was Copernicus a terribly bright astronomer; he did not make accurate observations. We have a few of his observations and they are mostly wrong and his tables are not particularly good either. But Copernicus had an idea which was very characteristic of a Renaissance revolutionary thinker. It was that the old system of the heavens was really a mess. One had only to try and read Ptolemy and consider all those eccentrics and epicycles to realise that nature could not make such a mess as that, it must have been men's thinking that had made such a muddle. He wanted to straighten it out. Fig. 55 is taken from Copernicus's book. It illustrates something we know now. The circles were added as an 'extra' but everyone liked circles then, ellipses were not to be admitted until the next stage. The sun is plainly there in the middle – with Mercury, Venus, the earth with the moon round it, Mars, Jupiter and the fixed stars, as the new immobile sphere.

You must remember that the old sphere was supposed to get round that enormous distance every twenty-four hours and a great deal of celestial mechanism was required for that. But this sphere was easy, you only had to move a few planets about. This was what Copernicus himself had to say about it:

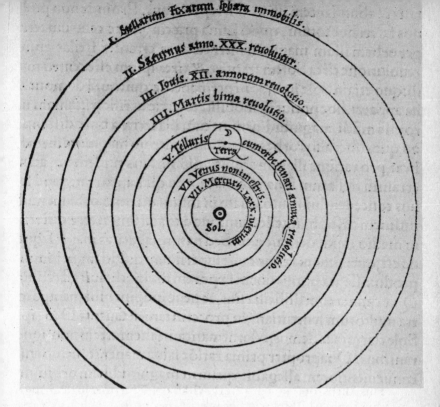

55 The Universe as illustrated by Copernicus in *De revolutionibus orbium coelestium*, 1543: I Sphere of the Fixed Stars; II Orbit of Saturn; III Orbit of Jupiter; IV Orbit of Mars; V Orbit of the Earth and Moon; VI Orbit of Venus; VII Orbit of Mercury; the sun is in the centre

Our ancestors assumed . . . a large number of celestial spheres for this reason especially, to explain the apparent motion of the planets by the principle of regularity. For they thought it altogether absurd that a heavenly body, which is a perfect sphere, should not always move uniformly. They thought by connecting and combining regular motions in various ways they could make any body move to any position. . . .

You note that this is a purely mathematical question, what is called 'saving the appearances'.

Therefore it seemed better to employ eccentrics and epicycles, a system most scholars finally accepted. Yet the planetary theory of Ptolemy and most other astronomers, although consistent with the numerical data, seemed likewise to present no small difficulty. For these theories were not adequate unless certain equants were also conceived; it then appeared that a planet moved with uniform velocity neither on its deferent nor about the centre of its epicycle. Hence a system of this sort seemed neither sufficiently absolute nor sufficiently pleasing to the mind.

Having become aware of these defects, I often considered whether there could perhaps be found a more reasonable arrangement of circles, for which every apparent inequality would be derived and in which everything would move uniformly about its proper centre, as the rule of absolute motion requires. After I had addressed myself to this very difficult and almost insoluble problem, the suggestion at length came to me how it could be solved with fewer and much simpler constructions than were formerly used, if some assumptions (which are called axioms) were granted me. They follow in this order. 1. There is no one centre of all the celestial circles. 2. The centre of the earth is not the centre of the universe, but only of gravity and of the lunar sphere. 3. All the spheres revolve about the sun as their mid-point, and therefore the sun is the centre of the Universe. 4. The ratio of the earth's distance from the sun to the height of the firmament [that is, the zone of the fixed star] is so much smaller than the ratio of the earth's radius to its distance from the sun that the distance from the earth to the sun is imperceptible in comparison with the height of the firmament.

This turns out to be a very vital point in theological and philosophical thinking because it makes the universe very much bigger. There is nothing absolutely new in this in another sense, because I have quoted a similar statement to you before, not from Copernicus but from Archimedes referring to Aristarchus. But Archimedes had now been translated and a scholar like Copernicus had read the passage which we considered earlier about the size of the universe. Further on, he states that the earth has three motions. First, it revolves annually in a great circle about the sun. Second, it revolves on its axis. The third motion is due to the variation in the inclination of the earth's axis to the line joining the sun and the earth. On account of the distance of the sun from the centre of the circle, the apparent motion of the sun is not uniform. Now, to explain this lack of uniformity was one of the greatest problems and I wish I had space to deal with it in more detail. Most people do not worry about it, but astronomers had to worry about it. How is it that the summer is not the same length as the winter; that the equinoxes do not occur half way through the year, that they are not half way between the solstices? The explanation was that the earth, as we would say now, or the sun, as they said then, moved slower at one part of the year than another. There had to be a lot of fiddling, so to speak, to get that to come right. This was the fiddling which Kepler did. But the really vital explanation was given, as above, by Copernicus and was another justification for his describing the planetary orbs as follows:

In the middle of all sits sun enthroned. In this most beautiful temple, could we place this luminary in any better position from which he can illuminate the whole at once? He is rightly called the Lamp, the Mind, the Ruler of the Universe; Hermes Trismegistus names him the Visible God, Sophocles' Electra calls him the All-seeing. So the sun sits upon a royal throne ruling his children the planets which circle round him. The earth has the moon at her service. As Aristotle says, in his *de Animalibus*, the moon has the closest relationship with the earth. Meanwhile the earth conceives by the sun, and becomes pregnant with an annual rebirth.

Now, this is the real stuff! I mean it is magic, it is analogy, and yet it happens to be true. But you can see that the reasons which give rise to an idea are not necessarily the same reasons as will prove that the idea is correct. Copernicus wrote all this down in his *On the Revolutions of the Celestial Orbs*, usually referred to as *De Revolutionibus*. The orbs were the same, he still thought of the planets as being fixed on the surfaces of a set of hollow crystal balls, but he had made a piece of celestial clock-work which was much simpler. It was also much bigger – and that was to cause trouble later. He himself succeeded in this enterprise remarkably well. He wrote the book, he presented it to the Pope, the Pope approved of it and it was circulated among discerning people and printed. It is true that this was already at the time when heretics were around and it was printed by a Protestant printer, but it was not disapproved of by the Pope and it stayed perfectly respectable right down to 1615, that is, for nearly a hundred years, before anyone really said there was anything seriously wrong with it.

So, you see, one might first of all arrive at a new answer, but then reasons have to be found for the answer; then the criticisms – which were many – had to be overcome; and that will be the subject of my next chapter.

7

Science and Religion

In my last chapter I was dealing with the first stage of the Scientific Revolution, the stage where the old ideas of the movement of the heavens were being attacked and an entirely different idea, if not an entirely new one, was being put forward by Copernicus. That was in the middle of the sixteenth century, around 1540. This idea was taken up by certain other astronomers and, on the basis of the Copernican picture, another set of astronomical tables was drawn up, the so-called Prussian Tables, indicating, as it were, where the centre of science was moving. Prussia, in those days, extended further east than it does now, and the area where Copernicus worked was called alternately Prussia or Poland. Astronomers still went on using the old methods, but they had to work much more accurately if their work was to be of practical use in navigation.

Tycho Brahe

The older observations were perforce limited because of the scale of the apparatus available and because of the inaccuracies of the observers. One of the maritime nations, a kind of half-way nation, Denmark, happened to be enjoying then a period of extreme prosperity. It had become prosperous by taking tolls on all the traffic going in and out of the Baltic through the Sound. The reigning king of Denmark was Frederick II and one of his noblemen, Tycho Brahe, who had the privilege of a good slice out of the toll of the Baltic, was also given an island in the Sound on which he set up what was the first scientific institution of modern times, called Uraniborg or Heaven's Castle (Fig. 56).

Unfortunately, Tycho was rather a hard man. Apart from being a good scientist, he was very quarrelsome. He became

56 Tycho Brahe's Observatory at Uraniborg

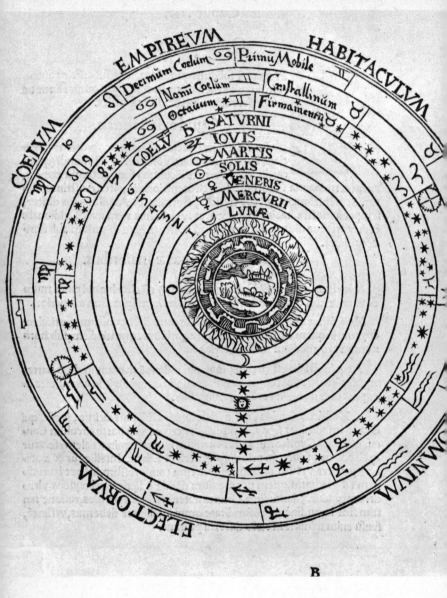

57 The old Aristotelian cosmology as illustrated in Gemma Phrysius's revision, *Cosmographia*, 1539, of Peter Apian's *Cosmographius Liber*, 1524

involved in a duel, had his nose cut off and had to wear a silver nose for the rest of his life. He was very hard on his tenants, so that the moment he was out of Uraniborg, the local peasantry tore it down and took everything away. There was absolutely nothing left for subsequent historians. We have only this picture of it which shows what an observatory was like in those days. It was really a typical nobleman's house with a few special scientific additions. The observatory part had four places from which to watch the stars, and another two higher up. The rest were Tycho Brahe's own apartments except for the basement which was entirely devoted to laboratories. The chemical laboratories can be clearly seen. All kinds of sciences were carried on together there and Tycho had, as I have said, a considerable amount of money with which to do so. After some difficulty, he was able to get very useful assistance.

I will now explain the revolution in the understanding of the world that was being effected. Fig. 57 shows the old cosmology: at the centre there is the earth with the animals and plants on it and the elements: earth, water, air and fire; and then the spheres and the moon, the sun, etc. Then, outside the zodiac, at the top, the empyrean or seventh heaven, the sky, the dwelling place of God and all the elect – a very nice, closed world. Fig. 57 is similar to Fig. 55 (p. 157), but with a different centre, that is all.

Tycho Brahe knew of Copernicus's work but did not choose to follow it all the way. He made a translation of it which preserved the central position of the earth but left everything else where Copernicus had put it, that is, he said the moon and sun went round the earth but the planets went round the sun. This made sense geometrically but not physically, because it still required the whole system to circle the earth once a day.

However, Tycho Brahe was more concerned with getting very accurate measurements and for that he had a set of instruments which were not very different in principle from earlier ones but they were larger. The observations were repeated with them more often and carried on for a longer time and the results were extremely detailed. He had trouble in getting suitable assistants for this particular work but he managed to get hold of a rather wild young German called Johann Kepler who worked with him for a few years. In 1588, King Frederick died and Protectors were elected to govern

Denmark during the minority of the young Prince (Christian IV). Soon afterwards Tycho Brahe, who had always been disliked by his fellow aristocrats, began to lose favour with the court, which had never been much interested in his work. In 1596 he was deprived of his appointments and of King Christian's protection. The next year he left Uraniborg. He brought the whole observatory to a standstill and transferred the instruments elsewhere. In 1599, he became chief mathematicus of the Emperor of Austria, Rudolph II. This was a very queer Emperor who never went to Vienna but lived all his time in Prague. He set up not only an astronomical observatory but remarkable alchemical laboratories that are still in existence and he harried his mathematicus to provide him with all the necessary horoscopes.

Meanwhile, Tycho Brahe left all his observations and records – there was a great deal of trouble about this – to his assistant Kepler. Kepler had great difficulty in getting them away from the family, because in those days they still believed that the only possible successor to an astronomer was obviously his son and the accumulated records should all be kept in the family. But Tycho Brahe's sons were almost incapable of mathematics, much less of astronomy and they agreed to hand everything over – I think for some consideration – to Kepler. It did not always happen like that. For instance, in France the astronomer Cassini was followed by his son Cassini II, then Cassini III and Cassini IV. For more than 120 years, they were hereditary astronomers of the Kings of France and they were, as it happens, all very good astronomers.

Johann Kepler

The fundamental change, however, was to be effected, not by Tycho Brahe, except that he was absolutely essential in providing the observations, but by Kepler who interpreted them. Kepler's main purpose was to make sense of the picture. He could see that the planets went round, he could even venture measurements based on the Copernican system, which were possible by that time although not previously. Mutual distances could be measured and one could see exactly how the planets did go round. Kepler was able to sense that the whole thing must fit into a system of some sort, that it must have some kind

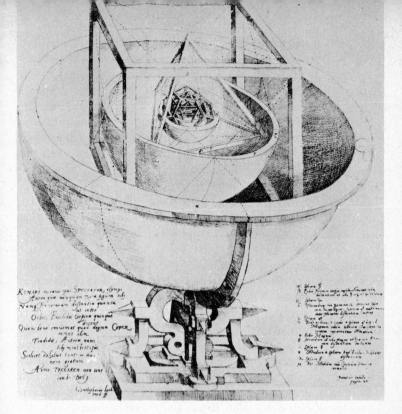

58 Kepler's first attempt to interpret planetary motion; from *De Harmonice Mundi*, Linz, 1619

of harmony. His first effort on it was typically along the Greek lines (Fig. 58). He believed that the successive orbits of the different planets were arranged so as to fit a set of polyhedra, one inscribed inside the other. There were five planets, there were five regular solids. It seemed obvious that the five planets must be related to the five regular solids, just as the Greeks had considered that the five elements were related to the five regular solids – a kind of analogical logic. In the centre there was a icosahedron, around the icosahedron a pentagonal dodecahedron, around that a tetrahedron, around that a cube, the cube fits into a sphere – and there you have the whole system. It was not so very far wrong, but the curious thing about Kepler's system was that despite this it was wrong to a degree that the experiments or the observations would not allow. Nevertheless, Kepler became the father of modern theoretical as well as experimental physics because he recognised that, while you had perfect freedom in working out your theories, they had to fit the observations. Now Plato and other people produced theories but when they found that the theories did not fit the observations, they merely thought it so much the worse for the observations! In effect, they said, 'there is an ideal world, and if the real world is bad, well, we must expect everything to be bad in a real world, so we will stick to ideal astronomy'.

But Kepler was not satisfied with this attitude. He spent a considerable amount of time and trouble over it and wrote a long book about it which he called *The Harmony of the World*, 1619. Finally, however, he gave the whole thing up because he could not make theory fit the observations. Then he went back and took a second look at them, giving himself a great deal more trouble. The second look was to try and solve the problem of why the planets did not go round regularly, why they went round faster at some times and slower at others. This had been explained in the old system by putting in such things as eccentrics and deferents and other purely mathematical tricks to save what are called appearances. You can always save appearances if you introduce sufficient constants! A new period of revolution could always be compiled out of a number of sub-periods.

But Kepler wanted something simple and, in a sense, he got it but by going a very long way round. Fig. 59 is taken from his book *Commentaries on the Motion of Mars* (1609). He concentrated on this one planet and, of course, it was the most appropriate planet on which to concentrate because it is the nearest external planet as well as one which can be observed at all times; it has a period of about 2:3 in ratio to that of the earth. Kepler tried various curves for its orbit and none would fit. Conventional epicycles would fit but they were not quite correct, they were eight minutes of arc out. Now, in the past, eight minutes of arc would not have mattered – no one could measure accurately enough to eight minutes of arc. The sun's image would need to be divided into quite a number of parts because the sun is thirty minutes of arc. But Kepler's measurements – or, rather, Brahe's measurements – were too accurate to admit an error as great as eight minutes of arc. If Kepler had not

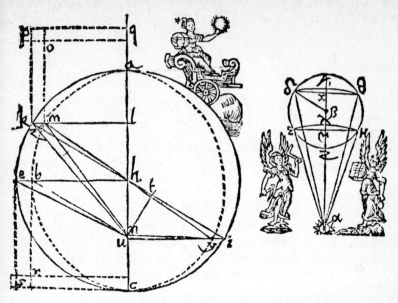

59 The elliptical orbit of Mars in its chariot around the sun; from Kepler, *Astronomica Nova . . . de Motibus Stellae Martis*, Prague, 1609

come until later, people would have continued to make measurements and make them very accurately without any theory – there is no reason why there should be a theory although it helps in a way. It is possible to make more and more accurate observations without any theories at all. However, if he had made more accurate observations, Kepler would not have succeeded, because an elliptical orbit fitted the degree of accuracy of his observations.

He decided that the planet Mars travelled in its chariot around the sun in an ellipse. In Fig. 59 you can see the ellipse which he developed as a certain proportion of the corresponding circle, and the sun is shown at one focus. Out of this he derived his first Law. This is a straight observational Law: he had no idea that it could make any sense dynamically because there was no knowledge then of dynamics. But, observationally, it was a simple curve – a curve which had been studied by the Greeks – and next to a circle it was the simplest of concepts. So he stated that all planets must travel in elliptical orbits of which the sun is the focus and he went on, of course, to show that the moon travelled in an elliptical orbit around the earth.

That was Kepler's First Law, which pleased him very much for quite a long time, but he was impelled to look further.

At that point the idea of an elliptical orbit, with the sun as one centre, immediately raised the question of how rapidly the planet moved round in orbit. Here he showed that it did not move round at a uniform speed, but faster when the planet was near the sun and slower when it was further away. From that he deduced that the actual speed it moved was such that it described the equal areas in equal times. It at once solved the problem of the inequalities of the times taken in different seasons – why, for instance, the summer, when the earth is far away from the sun, was longer than the winter, when it was closer. This may seem strange, that summer in the Northern Hemisphere was hotter, not because the sun was nearer but because the earth was exposed to the sun at higher altitudes and for longer days. Effectively, all that the Second Law meant was that the momentum of the earth in its orbit was constant. In other words, the product of the velocity and the radius vector remained the same.

So far Kepler had worked out from sheer observation the law of movement of a single planet, but to cover all the planets together, he needed to know the relation of the periods of rotation of the planets and their distance from the sun. This, he concluded after many years of work, was in the ratio of the square to the cube. In this way he finished the complete description of the motion of the planets in quantitative terms. His solutions, in fact, contained all the material necessary to work out the laws of force of the planetary system and made sense of Copernicus's intuitions. But before this could be accepted, the possibility of the system had to be established by direct observation. This was to be the work of Galileo Galilei.

Galileo Galilei

Galileo was the Professor of Mathematics and Military Engineering at Pisa which was the University town of Florence, rather as Cambridge was of London, and he was later transferred to Padua which was the University town of Venice. He was a man with very powerful connections, he dealt with princes and popes and he was a very good and popular lecturer. He covered a very large field of science. The interesting thing to note is that in those days scientists were not as limited as they are now. My point here is that Galileo was not an astronomer. His contribution consisted essentially of one set of observations, which were absolutely crucial, of the moons of the planets Jupiter and Saturn, and of the much more fundamental and far-reaching ideas leading to the revolution in dynamics. But for the moment I will deal with the first contribution. This is what Galileo himself says in the preface to a book published in 1610 entitled *Siderius Nuntius* (Messenger from the Stars).

... unfolding great and marvellous sights and proposing them to the attention of everyone, but especially of philosophers and astronomers, being such as has been observed by Galileo Galilei, a gentleman of Florence, Professor of Mathematics in the University of Padua, with the aid of a telescope lately invented by him ... has inspected the Moon's surface and an innumerable number of fixed stars, the Milky Way, the nebular stars, and especially respecting four planets which revolve about the planet Jupiter at different distances and at different periodic times with amazing velocity, and which after remaining unknown to everyone up to this day the author recently discovered and determined to name the 'Medicean Stars'.

Now, that is really good scientific publicity. What is even more, he had it published much more quickly than we could any of our papers – he was able to get it published within, I think, a month of having made the discovery. At that time, all one had to do was to go to the printer, get him to set up the type, watch him print, correct the proofs, and that was that. The book was spread all around the learned world within a year and it created a sensation – even our poet Donne was very excited about it. He wrote an essay called 'Ignatius his Conclave' about a new world in the moon – of which more later.

To go back to Galileo; his book opens with the words, 'About ten months ago a report reached my ears that a Dutchman had constructed a telescope'. Now, the Dutchman is usually referred to as Lippershey but no one knows anything about him and, of course, this telescope was a military secret because at the time the Dutch were occupied in wars against the Spaniards and Duke Maurice of Nassau had probably been using the telescope for some time. However, it got through by word of mouth that a telescope was being made and Galileo received confirmation of the report in a letter written from

Paris. I will not discuss the telescope itself here as I will be dealing with it in a later chapter on optics. But to continue with Galileo's account: 'After a little time I succeeded in making such an instrument through deep study of the theory of refraction.' And it had to be pretty deep because Snell's Law of refraction was not known at the time, it was only discovered some ten years later. However, the instrument worked, which was the essential thing.

And then bringing my eye to the concave lens I saw objects satisfactorily large and near. It would be altogether a waste of time to enumerate the number and importance of the benefits that this instrument may be expected to confer when used by land or sea. But without paying any attention to its use for terrestrial objects I betook myself towards observations of the heavenly bodies.

It is not quite true that he took no notice of terrestrial objects: he took a great deal of notice and did not do too badly out of it either. He presented the instrument to the Signoria of Venice as being extremely useful in naval warfare, enabling the number of ships and vessels of the enemy fleet to be counted before they could count yours, so that there was time to decide whether to fight or run away. He offered it to the Signoria free, gratis and for nothing and received a little consideration in the form of 500 scudi for it. This was a bit of luck because, as we can learn from Brecht's play on Galileo, within a year Dutch telescopes were selling in Venice for a couple of scudi apiece. The point was that he got in quickly – but he did the same on the astronomical side too. He looked at the moon and noted all the ruggedness of its surface but more important was the fact that he saw that there were far more stars than anyone had seen previously – to the naked eye there are only about 2,000 stars. If you look at the sky at night you may think that you see a great many more but, in fact, you can only, as I say, count 2,000 with the naked eye.

Galileo drew pictures of extra stars and on 7 January of the following year, 1610, he said:

There remains a matter which seems to me deserved to be considered the most important in this work, namely, I should disclose and publish to the world the occasion of discovering and observing four planets, never seen from the beginning of the world up to our time, their position and observations being made during the last two months about their movements and changes of magnitude, and I summon all astronomers to apply themselves to examine and determine their periodic time which it has not been permitted to me to achieve up to this day owing to the restriction of my time.

He worked too quickly, you see, but he wanted the publicity all the same! 'I give them warning, however, that they may not approach such an enquiry to no purpose, they will need an accurate telescope.' The Professors of Astronomy of the Universities of Italy said that as these stars could not be seen in the ordinary way and had not been noticed by Aristotle, they could not yield any useful information and they, themselves, did not want any telescopes. (I may say that at a certain college not too far away from here, the same thing was said not so very long ago about the electron microscope, that they did not see any use in having an electron microscope because good biologists knew everything that could be learned from an electron microscope without possessing one.) At any rate, they refused to look through them so that at first Galileo had no trouble with competition. The essential thing, to return to Galileo's account, is that:

On the 7th day of January in the present year, the first hour of the following night, when I was viewing the constellations of the heavens through a telescope, the planet Jupiter presented itself to view. I noticed a circumstance which I had never been able to notice before ... namely, three little stars, small but very bright, which were near the planet, and though I believed them to belong to a number of fixed stars, they made me somewhat wonder, because they seemed to be arranged exactly in one straight line, parallel to the ecliptic, and to be brighter than the rest of the stars equal to them in magnitude. The position of them with reference to each other was like this: two on one side and one on the other. Now I am scarcely worried at all by the distances between them and Jupiter when I first believed them to be fixed stars, but when on January 8th, led by some fatality, I looked again at the same part of the heavens I found a very different state of affairs. For now there were three little stars all to the west of Jupiter and nearer to it than on the previous night, they were separated from one another by equal intervals. At this point, though I had not turned my thoughts at all to the approximation of the stars to one another, yet my surprise began to be excited, how Jupiter could one day be found to be to the east of the same stars when the day before it had been to the west of two of them. And forthwith I began to be afraid that the planet might have moved differently from the calculations of the astronomers.

Because, naturally, he still looked upon them as stars.

And so it passed the stars by no proper motion [i.e. going the wrong way]. I therefore waited till the next night with the most intense longing, but I was disappointed, the night was covered with cloud in every direction. But on January 10th the stars appeared in the following positions: two on one side and the third could not be seen at all, as I thought being hidden by the planet. When I had seen these phenomena I knew that the corresponding changes in the position could by no means belong to Jupiter. Accordingly on January 11th again they were both on the east. I therefore concluded and decided unhesitatingly that these three stars in the heavens were moving about Jupiter, and Venus and Mercury about the Sun which was at length established as clear as daylight by numerous and subsequent observations.

You see, here is what you might call the actual observational proof, which does not require any mathematics, but which does require observation. Here we have a model of the Copernican solar system and this was what really convinced people, not the calculations of the astronomers, not Kepler's laws, but the straightforward observation. Of course, Galileo saw the significance of it; he realised that these stars in the heavens were really a celestial clock that gave absolute time and, incidentally, something else too. With absolute time he realised that he had discovered just what was needed by the navigators – a way of finding the longitudes. All that was required was to set up a nice table of Jupiter's satellites – which was prepared later in the nautical almanacs – look up the times of Jupiter's satellites, at the absolute time, then look at the local time and you established the longitude. It all depended on having good enough tables of Jupiter's satellites. Galileo thought this was a wonderful and most useful idea and he proceeded to try to sell it.

He needed money as he had two daughters whom he had great difficulty in marrying off and in the end, in fact, he had to have one of them put in a convent. He was really very short of money as his salary was not sufficient. In a letter to the

King of Spain, Galileo offered this way of finding the longitude, pointing out that finding longitude was of no use to him as a private astronomer but that the King of Spain, with his wide dominion, could certainly use it to very great advantage. He also added that the King of Spain, being an absolute monarch and being able to decide things for himself, would be able to understand its importance and would immediately make use of his offer. He had, of course, offered only the names of the new stars to his previous patrons and called them the Medicean stars. But the Medici were mean, they did not provide money so he withdrew the name Medicean Stars and called them by purely classical names.

The offer to the King of Spain was a very practical one. Very unfairly, the King referred the whole matter to a committee. The committee sat on it for about three years and then decided that it was not in fact a very practical method – the sailors could not observe the satellites of Jupiter from a ship – and the offer was turned down. But Galileo was not put off by that; a few years later he adapted his letter and sent it to the States General of Holland, pointing out that as practical men, being democrats and not liable to the autocratic rule of one person, and so forth, they would immediately see for themselves how advantageous this method was. Unfortunately, the States General of Holland did not refuse it, they simply held on to it and would not pay anything to Galileo for it. In any case, the method did not work – in fact it never has worked. Why it did not is a very nice point involving the velocity of light but, in fact, this method was never much use in navigation.

The foundation of dynamics

Thus, we have the solar system established, but now comes the question of the other aspect of the theory, to which I referred earlier – the foundation of dynamics. I will go back a little. The real foundation of modern dynamics was to be found in the cannon ball. Aristotle had some very good ideas about dynamics: everything fell to its natural place; the natural place of everything was according to the element; if it was an earthly thing, a solid thing, it fell to the ground or right through the sea to the bottom; if it was fire it went upwards. But what about what is called 'violent motion'? What about throwing a stone or shooting an arrow? Well, it was very difficult to explain that on Aristotelian grounds, so the theory was that natural motion was all right – the stone fell, and the heavier the stone the faster it fell – but violent motion was much more difficult to explain. Aristotle argued this way: you threw a spear, the spear penetrated the air, the air which was displaced from the front end of the spear came round to the back end of the spear and pushed it along. It did not do this for very long because it tired of it and, ultimately, the spear always fell to the ground. This satisfied most people for about 2,000 years, though some did have doubts about it.

One of the first people to express such doubts was a philosopher of the sixth century AD called John Philiponos, who was converted to Christianity. The first premise of Christianity in those times, for very obvious reasons, was that everything Aristotle had said was wrong. Incidentally, 600 years later, Christianity had come round to accepting Aristotle and insisting that everything he had said was right; and it was only in the Reformation that they again came back to saying 'everything Aristotle had said was wrong'. This was the title of the thesis that Ramus offered to the University of Paris; later he was killed in the massacre of St Bartholomew. But John Philiponos, in his time, had said, no, Aristotle's theory about the spear was not the case at all. When you threw a stone it had something in it called impetus, and the impetus carried the stone along and as long as it preserved its impetus the stone would go on, when it lost its impetus it would fall. The impetus trajectory, therefore, was something which went along in the air and then came down – and that was very much like what cannon balls did.

With a siege cannon, you fired it so that the stone went up and over the wall and fell down on the other side, or you shot it at the wall at point blank range and it breached the wall. That, however, was about as much as could be got out of a primitive cannon, apart from frightening the horses! The cannons were used at the battle of Crécy and, partly as a result, the French cavalry were thrown into disorder; this was reported in the first account of Crécy, as written by Froissart, the French historian. But, afterwards, as the English were winning the war and he had transferred his services to the English side, he rewrote the account of the battle omitting all account of the cannons because they were considered unsporting. So, if you read the English version and not the French version you will probably never hear about cannons being used at the battle of Crécy.

Fig. 60 shows how the cannon were made with a special boring tool, but they were never very accurate. The cannon balls were even less accurate; they rolled around in the barrel and there was what was called windage, so that the theory was not really much use. In fact, it was of no use at all until the development of anti-aircraft gunnery in the Second World War. Boyle, I think, said: 'The cannon has made more noise in philosophy than it has in warfare.'

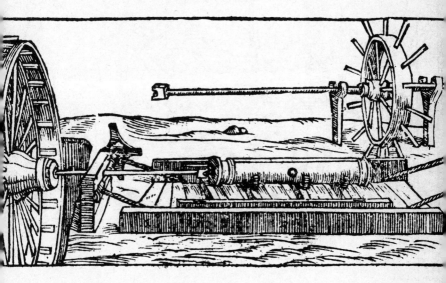

60 Drilling cannon; from Biringuccio,
De la Pirotechnia

61 The art of the gunner: taking aim; from W. Ruff, *Der . . .mathematischen und mechanischen Kunst eygentlicher bericht,* Nuremburg, 1547

Fig. 61 is concerned with the gunner's art; the gunner is here taking aim and he can, with wedges, cock the gun up or down as well as fire it at point blank. The quadrant was put into the barrel of the gun and the angle of elevation read by means of the plumb line.

Fig. 62 is taken from a Spanish book published slightly later, and shows the ballistics of the times. This is not quite strictly based on the impetus theory. People had realised that the ball did not fall straight down but, according to the inclination, it first had a straight piece of trajectory – that was where it had its impetus – then it curved round and came plumb down. As you can see, the last part of the curve is quite a good free-hand approximation to a parabola – of course the trajectory is not a parabola at these low velocities – but it is clear that they had grasped the principle of the thing. If the gun was firing level or at point blank, it did not have much range. If you raised it a little the range increased. If it was put up to 45° it acquired a bigger range still and if it was put up beyond 45° it had a smaller range again. This was what started modern ballistics.

Fig. 63 illustrates a very elaborate concept that encompasses

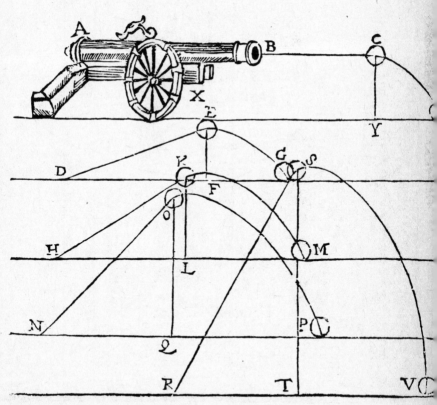

62 The study of ballistics; from Cespedes, *Instrumentos Nuevos de Giometria*, 1606

63 The concept of New Science; from Nicolo Tartaglia, *Nova Scientia*, Venice, 1537

the whole of science. It is by a character called Tartaglia, who was quite a good mathematician, and he is pointing out that this is the way to learning. You have to go through the door marked mathematics, with Euclid guarding it. Inside, apart from a whole range of sciences – mathematics, music, astronomy and all the muses are awaiting you – two very interesting things can be seen: a gun which is firing a low trajectory and a mortar which is firing a high one. Now, Tartaglia says in his book on ballistics: 'I was present at a dispute between gunners on the subject of what angle you must put the gun in order to get the greatest range, and they took bets on it. I gave this very careful mathematical consideration and worked out that the maximum angle must be 45°.' This is certainly a very good mathematical answer – it is more or less correct neglecting the air resistance – but his method of arriving at it was, we might say, almost quantum. If you fire the gun horizontally, in principle it has no range at all; if you fire it vertically, it comes back on top of you. Therefore, if you halve the angle, that must be the maximum. However, although Tartaglia wrote his book on ballistics, or, rather, he did all his work on ballistics, he did not publish the book straight away. He became conscience sticken by what he had done – this is the first appearance of this problem in science – he had found means of perfecting weapons of destruction which would be used 'in the numerous wars which Christians were fighting against each other'. Was this a thing he could publish with proper concern for his soul? He decided it was not. So he burnt all his notes and did not proceed with the publication. But then, unfortunately, Italy was threatened by an invasion from the Turks, who were infidels, and in that case the Christian principles did not hold. So he did his calculations again and published the book. Now, although Tartaglia started this work, he did not get very far with it and it was taken up later by Galileo.

The laws of falling bodies

This, then, is the other side of Galileo's activities, the side of arguing about the motion of bodies: cannon balls or falling bodies, natural motion or artificial motion. It was at that point that Galileo began a set of experiments in order to find out what the laws of falling bodies really were. We are told, of course, that he dropped two weights of different mass from the leaning tower of Pisa and found that they reached the bottom at the same time. This experiment has a curious history, but one part of it is fairly certain – that such a thing was never done by Galileo. It was done by a monk a little after Galileo's time to verify the experiment that Galileo never carried out and, curiously enough, it is now being used in the National Physical Laboratory as the most accurate way of determining the value of the acceleration due to gravity, g. For this, you have a little sphere – a little quartz marble – which is flicked up and the time measured that it takes to go up as well as the time it takes to come down again – exactly, of course – and from that you measure g. It is more accurate than any pendulum method, but it is very tricky. Very good results have not been obtained yet because the wretched thing will not always fly straight up and it does not always come back to the same place. The National Physical Laboratory have not really got the 'bugs' out of it yet. But, in any case, although Galileo did not actually carry out this experiment, he carried out a number of thought experiments of the same kind, such as imagining something falling which was made of two pieces and showing that the separate pieces must fall at the same speed as if they were together. Therefore, all things fall at this speed – neglecting the resistance of air. He consciously neglected the resistance of air, because he knew that there was air resistance. He was trying to work out, and this is another example of a kind of idealism in theoretical physics, what bodies would do in free space or in vacuum.

When I was talking about Aristotle earlier, I omitted to mention a very important point. Aristotle's theory of motion is, of course, impossible except in some air, because the air has to be there so as to come round to the back (of the spear); if there were no air it could not work. Therefore, said Aristotle, the vacuum is impossible. But he knew a bit more than that. He said, if there were a vacuum any body would continue its motion at uniform velocity in a straight line – which is absurd; therefore there cannot be a vacuum. Aristotle had enunciated precisely what became Newton's First Law of Motion and he used it as an argument against the existence of a vacuum. Which shows that you have to be very careful about saying things that are absurd. Some things are absurd, of course, things that could not be done here in this room. This book can be pushed along a table but it will not go on moving in a straight line for ever. Also, a cart on a level surface must have a horse to pull it along, as everyone knows. Aristotle's dynamics was much more akin to real life, but care must be taken about using notions which are common notions without understanding their implications.

However, Galileo reasoned this out and, further, he worked out the actual laws of fall, that is, a natural motion in the Aristotelian sense. Everyone knew that as you dropped something it went faster the further down it went – even Aristotle knew that. But Aristotle said: 'It is trying to get to the ground: just as a horse moves faster when it gets near to its stable, so the stone moves faster, the nearer it gets to the ground.' That is what is called the doctrine of final causes, that is, things are determined by what they are going to be rather than what went before – and in its place it is an extremely good doctrine. The trouble about Aristotle was that he was a biologist. The above is a perfectly good doctrine for horses but it is not any good for inanimate objects. We do not use the concept of final causes in physics but it is very useful in biology.

So, to go back to Galileo's measurements. There has been a great deal of argument about these. As I mentioned earlier, Aristotle had been taken up again in the later Middle Ages and you get the impression that Aristotle ruled everything. But, of course, there was always an opposition and the great advantage of the early Catholic Church was that there were religious orders in it. The people who backed Aristotle were the Dominicans, among them notably the great Dominican theologian St Thomas Aquinas. But just because Thomas Aquinas was a Dominican, the Franciscans did not accept this at all and they put forward another philosopher Duns Scotus, and one of his followers Buridan, who argued the other way round on many other subjects, I am afraid, on almost every subject.

The opposition to Aristotle, therefore, was already to be found in schools like Oxford and Paris – they were not the orthodox schools but they were known and their views were argued about. The great thing about this argument was that it was conducted entirely on theological and philosophical grounds. Where Galileo differed from the others was that he

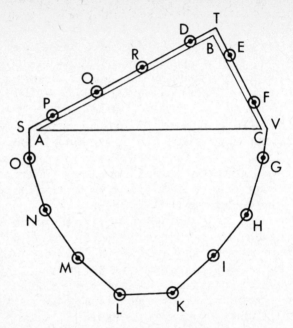

64 Frictionless chain hanging over a right-angled prism

conducted his arguments by experiments. He realised, however, that it was going to be extremely difficult, with the
apparatus he had, to measure actual acceleration of a freely
falling body. So he would avoid the problem by working on a
slope, where the acceleration would be proportionately less,
or by going back to his favourite pendulum, which he could
make move as slowly as he liked. The law of movement down
slopes had been discovered about the same time by Stevinus
of Bruges. It involved knowing how forces acting in different
directions could be compounded. The Greeks never did this.
Archimedes had always dealt with parallel forces in the laws
of the lever. But Stevinus hit on a very good example of
a thought experiment. Fig. 64 shows a frictionless chain
hanging over a smooth right-angled prism. A loop will not
move by itself: if it did, perpetual motion would be possible.
By imagining the chain cut off at the ends, it will still not move,
so the part down one side must balance the part down the other;
that is, force is inversely proportional to the length and, thus,
for a very low slope the force would be very low indeed and,

correspondingly, the acceleration would be small and therefore measurable.

Galileo made a special apparatus to deal with fall down a plane inclined at different angles. I tried to construct one for my lectures but it was never a success – Galileo was better off than I was in that he used wood while I tried to use steel. He used a parchment-covered v-groove and rolled a polished brass ball down it – and it was a really beautifully carried out experiment. His main concern was how to measure the time. He measured it by weighing it. He had a little equal-levelled water clock out of which came a very thin stream of water and when the ball started to move he put a vessel underneath, and when the ball came to the end of the slope or to a particular mark, he removed the vessel and weighed it. He then had no other means of measuring time. The first thing he found was that the distance traversed in corresponding times were as the square of the time, and from this he derived the law of motion. This was fundamental for dynamics and it was not known before.

He had earlier made the fundamental observation, which he could only understand later, about the pendulum. In an idle moment in the cathedral in Pisa, he noticed the swing of the various low lamps which were hung from the top of the dome. As they swung and then gradually came to rest, he wondered how it was that it seemed to him the lamp took the same time to swing whether it was doing a big swing or a little swing. But how could he find out? He could not use a stop watch because it did not exist then: this was, in fact, how the stop watch came into existence. But he had a clock of his own – his pulse. He counted the swings of the pendulum by his pulse and he found, within the accuracy of the method, that the time of the swing was independent of the amplitude of the pendulum. Then he went back and did it again with pendulums of different lengths and he found the relation between the length of the pendulum and its period: the square root law.

Galileo was a very practical person: he immediately invented an instrument which reversed the process, a very good thing to do. He had used his pulse to time the pendulum and the pendulum could now be used to time the pulse. Here, he thought, is an instrument he could give to the doctors. This was Galileo's pulsilogium. In Fig. 65 we can see that it occurs

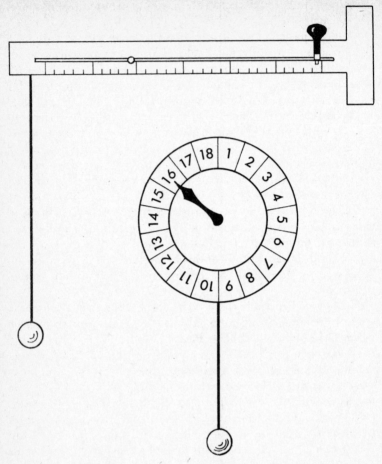

65 Galileo's pulsilogium

in two forms, both of which were basic instrumental points. The upper is a straightforward one: you adjust the length of the pendulum until its rate of oscillation coincides with the patient's pulse-beat and then you read off the period. No counting is needed. The other method is really to do the same thing but to have the string wound round a cylinder and read off its length on a dial. It can be calibrated. I do not think it was ever a great medical success because the ordinary doctor has the feel of this: he can tell without these instruments whether the pulse is a very fast one, a very slow one, a very weak one or a very strong one. It was like taking a temperature by just putting his hand on the patient to feel whether he was feverish or cold. For all practical purposes this procedure was just as good as taking a temperature then because no one knew – or knows today – what a temperature registered, what it ought to be or what any variation signified.

To come back to Galileo, he next worked out that the fall of a body at the end of a string in the swing of the pendulum was an example of free fall because the string did not affect it and he could work out a number of things in connection with the height. For instance, one experiment he did was to let a pendulum with a body at its end swing, then he caught the string on a nail and, of course, the body swung up: but he noted that it always swung up to the same height, apart from friction, that it had to start with though it was swinging on a smaller radius. He had arrived at the principle, therefore, that the energy of the body at the bottom of the swing was determined entirely and only by the height from which it had fallen and not by the path along which it had fallen. These were dynamic concepts which he discovered and which he described in his last work, called *Discourses on Two New Sciences*, dynamics and statics.

In this account of Galileo's work, I have not referred to what everybody quite rightly remembers Galileo for: his quarrel with the Church. This was provoked by his earlier *Dialogue on the Two Chief Systems of the World*, published in 1532 and presented to the Pope. Now, Copernicus had also written a book and it, too, had been presented to the Pope, and it was written very learnedly in Latin. But, not only did Galileo write his book in Italian, he wrote it not as a book but as a play, as a dialogue. Actually, it was not really a dialogue as there were actually three characters in it and they all argued with each other. There was the old Aristotelian called Simplicio, who was made a fool of all the time; there was Galileo himself as Sagredo, and one of those stooge friends who always say 'how true!' and so forth. It was, of course, aimed at the public and hit the public.

Galileo and the Inquisition

Unfortunately, the book was not very successful. Galileo had already created a virulent opposition to himself. People had preached against him and with all his various quarrels with the Church he had many enemies. True, he had many powerful friends, too, but in the end some spiteful person pointed out to the Pope that obviously the character Simplicio in Galileo's book was intended to represent the Pope – and this was going a bit too far. So Galileo was had up before the Inquisition who went a long way round to get to the point of the indictment. It was not an argument as to whether the earth went round the sun or the sun went round the earth; it was an argument about the theological consequences – because already there had been theological consequences of a very unfortunate kind. If the system of the world was not the old system of the world but the new system, first of all, where was all the outside region, Heaven, the empyrean sky where the elect lived and so forth? This would not fit into the new picture and that was bad enough. But what you *could* fit into the new picture was much worse – many other worlds on which there might be a great many people and were those people Christians? if not, why not? I do not know why people should worry about the religion of hypothetical people on hypothetical worlds, but they did. In about 1600, a philosopher called Giordano Bruno had got into real trouble over this. He had travelled round Europe fairly safely and came to England where he had long talks with our John Dee and others; but he would go back to Venice. This was unfortunate, politically, because, although normally Venice was a fairly safe place for him to be, as the Venetians were always quarrelling with the Pope, at that particular time they were not, and Bruno was handed over to the Inquisition and burnt alive. This naturally scared a number of people but, although Galileo was not scared, he was bullied and forced to recant, after which he was given a reasonable penance and sent to live in the palace of one of his nobleman friends where he wrote his books on dynamics.

The condemnation and recantation of Galileo was intended to be the end of the heliocentric solar system. It turned out, on the contrary, to be the beginning of its general acceptance – the end turned out to be the beginning. Galileo's defeat was supposed to have destroyed the theory of the rotation of the earth completely. He was condemned and even the men who supported him had to back down on it. But, of course, it was only then, as I say, that the idea really started spreading. An example of a real prophecy made by someone, which really did come off, was given by our poet Donne, as I mentioned earlier, in his essay *Ignatius, his Conclave*, which proposed a journey to a colonisation of the moon. In this essay they were having a Cabinet meeting in Hell to decide that only the people who really caused trouble in the world could qualify as ministers in Hell – and Copernicus rose. Then Ignatius, who was acting as Prime Minister, objected very strongly to this and said he had no claim in Hell. Copernicus said: 'Well, all these people upset things on earth, I have upset them in the heavens, which is far worse.' 'Oh!', said Ignatius, 'it is only theory, there is nothing in it: astronomers can say it but it has no effect on ordinary people.' He then went on to say: 'But if this theory should cease to be an astronomical theory and be condemned by the Pope, then both you and the Pope who condemns you have got a place in Hell.' That essay was written in 1611 and the actual condemnation did not take place until 1638, so there is a case, written and printed, of a genuine prophecy.

8

The Birth of Dynamics

I now come to the last section of the central part of the Scientific Revolution, the final expansion of knowledge to take the place of what had been destroyed in the earlier parts. The first part, I recalled, was the destruction of the old earth-centred world picture. In the first stage, there was an elementary destruction of this picture which, in a sense, replaced the earth by the sun; the planets still went round in circles. In the second stage, which I discussed in my last lecture, that picture was refined and, at the same time, justified. The justification was given by Galileo. Galileo showed that, in the moons of Jupiter, you could have a real model of a solar system, you could see it and measure it yourself.

Part of the second stage is due to the work of Kepler. What Kepler did, essentially, was to refine the picture still further. He demonstrated planetary orbits as ellipses instead of circles and eccentrics. His Second Law, though he could not grasp its significance himself, effectively meant that the momentum of the earth in its orbit was constant (p. 168), or that the moment of momentum of a body without a force acting on it is constant.

This was seen in a rough kind of way by the people who criticised Aristotle in their concept of impetus. But it is a transition from the concept of impetus to the concept of what Newton was later to call inertia, or the quantity of motion, the velocity times the mass. This concept was gradually refined, first by Galileo who showed it in the fall of bodies, in the trajectories of shot and in the motion of the pendulum, and then still further by Newton.

Isaac Newton

In coming to Newton we have to move again into a different

world. The easiest thing to remember is that Newton was born in the same year that Galileo died, 1642; that is, a complete life-time separated them. But, as we shall see in a moment, they were much closer than that. In fact, as we now know, Newton's chief work comes directly from Galileo and not through any intermediary at all. But the laws of motion of bodies had first of all to be established. Now, Newton was one of those extremely remarkable men of whom it can be said that only one appears in every two or three centuries. And this quality was evident from the start. In later times – and I have been slightly guilty of this myself – there has been a tendency to say, 'Well, what *did* Newton do? It would have been done in any case by his contemporaries.' But, from the information we have had recently, it appears that Newton really arrived at the answers extraordinarily quickly. In fact, he had most of the answers by the time he was twenty-one, although he did not publish them until he was forty-two.

Newton grew up in surroundings in which science was already an accepted thing, whereas Copernicus and Kepler and, particularly, Galileo had to live in a world where science was fighting for its existence. Newton never had to do that. Yet, he was a curiously suspicious character. He was one of the few people who actually resigned from the Royal Society, and for a very simple reason: he did not want to meet the Fellows because they sometimes actually disagreed with him and contradicted him, which was more than he could tolerate. He was, however, persuaded to return and to be President for about twenty-five years in his old age.

He was one of those people who work almost entirely on their own. The great work he did was done in a forced vacation in 1665 when the plague was raging in London and he went back to the part of the country where he was born, to Woolsthorpe in Lincolnshire. He was actually the posthumous son of a small farmer, but with quite respectable connections. His uncle paid for his upbringing, sending him to grammar school at Woolsthorpe and from there he went, albeit in a very inferior position, to Cambridge where he did not do anything particularly distinguished but where he did make friends with Dr Barrow, the Professor of Mathematics there. Dr Barrow was a very good mathematician; he was not very interested in physics but he did lay the basic foundations of algebraic analysis.

66 A page taken from the correspondence of Isaac Newton

Memorand that before yo[u]
it was agreed betweene [us?]
named mr George [...]
[...]

Henry: shepward:

James [...]

william [...]

I should remark in passing that this period of the early seventeenth century was the period in which mathematics took the form which it has now. There was symbolic expression, the use of letters, the beginnings of a differential calculus, the working out of mathematical series, and a great many fundamental ideas of geometry that were introduced by Descartes – co-ordinate geometry, for instance. But most of all, for practical purposes, there was the introduction of logarithms and not only for ordinary multiplication but, even more usefully, for trigonometry. Therefore, it was rather like introducing a computer, and the whole business of astronomical calculations could be done easily and quickly. If this had all to be done by long multiplication – they did this and even worse in those days – you can see how difficult these calculations could be. The actual tools of logarithms and computations were now available.

In Fig. 66 you can see a critical page in the key document which came to light in about 1961 and which shows that Newton really did make discoveries he is supposed to have made. As you can see – there is a certain amount of long multiplication here – he did not grudge doing a great deal of work although he was not particularly neat in the way that he did it.

Scientific societies

By that time there was a considerable movement of a more social kind in the development of science. During the Civil War a rather strange thing happened. Not unexpectedly, the Fellows and Masters of the Oxford Colleges proved to be remarkably loyal to His Majesty, and when His Majesty was finally driven out of Oxford, his headquarters, they were found to be security risks, so to speak. So they were removed from all their positions and were replaced by more republican-minded Cambridge men. Then, for a short period of about ten years, science flourished at Oxford – and it flourished in a big way. The new scientists or virtuosi met in each other's rooms and discussed things. There were people like Boyle and his technical assistant, Hooke; there was a character not normally associated with science, Christopher Wren; there were other learned people much more important in their time than they are now, for instance, Willis and Wilkins, and there was a rather odd character, Thomas Sprat. Thomas Sprat wrote *The History of the Royal Society of London*, which is a most remarkable history because the Royal Society of London was only eight years old when he wrote it. But its knowledge and its members made it important enough to merit a history being written of it. Thomas Sprat said: 'The principal and most constant of them were Seth Ward, the present Lord Bishop of Exeter, Mr Boyle, Dr Wilkins, Sir William Petty [the founder of economics], Mr Mathew Wren, Dr Wallis [a mathematician], Dr Goddard, Dr Willis [another mathematician], Dr Theodore Haak, Dr Christopher Wren and Mr Hooke.' These are the people, then, who were meeting in their various rooms at Oxford to form a Society for the Pursuit of Natural Knowledge. This Society was not the first one of its kind; one of the first was in Italy, it was called the Accademia dei Lincei at Rome (1603–30). This Academy of the Linx produced reports on mathematics and natural science. The only parallel to this in modern times is a French mathematical society called Bourbaki which publishes anonymous mathematical papers; no one knows who Bourbaki is.

The idea of a scientific society started in Italy and then spread to one or two other countries. It was a time when people met in each other's rooms in France as well as in England and really studied experimental science. They were following the idea of someone who was not a scientist himself but who wrote more sense on science than anyone has written since, that is, Francis Bacon. Bacon had the idea of science, he did not actually carry out any experimental work (although his death resulted from exposure to the cold while carrying out an experiment, it was probably the only one he did), but he certainly had the idea of it. With this idea, he examined inventions. He was the man for assessing the place and value of experimental science as well as for classifying and organising research:

. . . it was plain that the good effects wrought by founders of cities, law-givers, fathers of the people, extirpers of tyrants and heroes of that class, extend but over narrow spaces and last for but short times; whereas the work of the Inventor, though a thing of less pomp and show, is felt everywhere and forever.

But above all, if a man could succeed, not in striking out some particular invention, however useful, but in kindling a light in Nature – a light which should in its very rising touch and illuminate all the border-regions that confine upon the circle of our present knowledge; and so spreading further and further should presently disclose and bring into sight all that is most secret and hidden in the world – that man (I thought) would be the benefactor indeed of the human race – the propagator of man's empire over the universe, the champion of liberty, the conqueror and subduer of necessities.

That was written before the two cultures separated. I mean that it would be difficult to find a scientist who could write like that today. At any rate, this idea took root and it was the inspiration of the Royal Society. The 'royal' part of it is a bit of a joke. After the fall of the Commonwealth, the various more-or-less republican scientists whose politics did not go very deep, rapidly rallied to the King's side and got into touch with the same old courtiers – some very queer characters, incidentally – Hobbes, the philosopher, Sir Kenelm Digby, who believed in sympathetic magic, and a few others – not remarkable scientists but good royalists. They got together and obtained the patronage of the King. It did not cost the King anything nor did he bother too much about it. He never attended any meetings of the Society and he never gave them a penny, but he did allow them to use his name and that, of course, was very important indeed – more important then than it is even now. Thus the Royal Society was founded.

King Charles II had a paymaster, Louis XIV of France, who kept him going financially so that he did not need to call parliament and raise taxes. Fig. 67 shows Louis XIV inspecting his own Royal Society, the Académie Royale des Arts et Sciences, all done in very great style. You can see the King, feathers and all, and the President of the Society. You will note the bits and pieces of apparatus that belonged to the science of that time, a big glass for burning up diamonds etc. Practical science is represented in the form of compounds and fortifications. Distilling apparatus, thermometers, various mechanical

67 Louis XIV visiting L'Académie Royale des Arts et Sciences

devices, some computing devices, an air pump and a few illustrations of natural history can be seen in the picture and round the sides. In the background, a portent of modern days, you can see the observatory – the observatory of Paris which is still there and still used – very nice gardens, a skeleton of a deer, flowers – everything required for a scientific Society. The difference between that Académie and the Royal Society, incidentally, was that it worked in the reverse way: the Academicians were paid, whereas the Fellows of the Royal Society had to pay to belong. They did not pay very well; they should have paid a shilling a week and they were often years behind in their subscriptions. In fact, the Society nearly went bankrupt a few years later. Poor Hooke, who had the task of producing two new important experiments every week, received no salary whatever. Nevertheless, he was promised he would receive a salary when the funds of the Society were sufficient to pay it. However, he managed to live by securing little jobs like surveying the City of London after the Great Fire. It was a shocking affair when he died, because they found in his effects a large chest full of gold which he did not leave to the Society, and the Society took this very hard. But, in the circumstances, although he had no family, I personally do not think it surprising.

Fig. 68 shows our own comparatively modest effort. There is the first President of the Royal Society of London, Lord Brouncker, who was not a bad mathematician but a better courtier; there is the real inspirer, Bacon; and there are relatively harmless little instruments belonging to the Society. And, of course, the bust of the Patron is there, being crowned by some goddess or other. For the things of interest in the picture, there is the telescope of the day, suspended by a mast, mostly hollow and very long. I shall have more to say about that later. Also of some interest is the gun shown here, with which Brouncker did some experiments on the velocity and penetration of bullets. This picture covered the whole of the activities of the Society at that time.

We now come to the more practical side of the things in which they were interested. Fig. 69 shows the first use of the

68 The frontispiece of Thomas Sprat's *History of the Royal Society*, 1667. It shows Lord Brouncker, the Society's first president (left), Charles II, its patron (centre) and Francis Bacon (right)

CAROLVS
II.
SOCIETATI
REGALIS
AVTHOR
&
PATRONVS

...ETATIS PRÆSES   ARTIVM INSTAVRATOR

...io in B.D.C.   Wenceslaus Hollar f. 1667

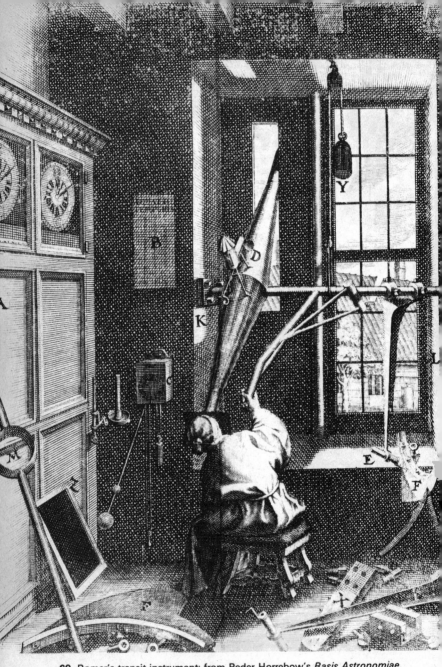

69 Romer's transit instrument; from Peder Horrebow's *Basis Astronomiae, Harniae*, 1735

70 The Royal Observatory, Greenwich, in the time of Flamsteed

telescope for astronomy. All the findings I have told you about so far were the result either of naked eye observations – as in the case of Tycho Brahe and Kepler – or by means of a hand telescope with the aid of which things could be seen but not measured. This figure illustrates the first transit instrument. You can see, it is just an ordinary room with a slip covering half a double window, the telescope and the clocks for measuring (pendulum clocks as well as rather longer term clocks) and all the bits and pieces belonging to the instrument. With it some really accurate measurements could be made.

Fig. 70 shows us the beginning of our scientific effort in this field, the Royal Observatory at Greenwich, now abandoned on account of air pollution but containing what was necessary for the times. This is a typical Dutch-style building. The King had something to do with having it erected. It seems there were some bricks left over from a palace in which a previous king had kept some mistresses, and the King graciously allowed these bricks to be used to build the Observatory which he had sited in his own royal park. And there it is to this day. You will note the main feature – a magnification of what you see in Fig. 69 – the long windows. Flamsteed was appointed Astronomer Royal, but he was not given much help. He had to buy

71 (overleaf) Interior of the Royal Observatory, Greenwich, in the time of Flamsteed, showing observers using the telescope and quadrant

the instruments out of his own pocket and he had to earn his own salary by teaching a number of naval apprentices. Nevertheless, he carried out his observations.

In Fig. 71 you can see the inside of the Observatory. There, very large and grand, you can see the portrait of the King and the Duke of York; the clocks, a quadrant, but this time with a lens in it, and the telescope. As you can see, the adjustment of the telescope was somewhat primitive. If you wanted to cock it up you had to put it through the next rung of the ladder. The hand adjustment at the lower end is for fine work. But observations could be made and, in fact, all the early observations were made there – if you examine the illustration you can see someone writing down the observations. This, then, was the beginning of modern astronomy.

They not only had telescopes – they had microscopes. Fig. 72 shows Hooke's microscope with which he made all his first observations of anything that took his notice, particularly such small animals as fleas and flies. He wrote a fascinating book on his discoveries of the cells in plants and he laid the whole foundation of modern biology. Nevertheless, he was not the best biologist of that period, the best was a Dutchman called Leeuwenhoek. Leeuwenhoek was a cloth merchant and he carried out his investigations for fun – he was one of the real amateurs. However, he could not be bothered to make a microscope like Hooke's; he used a simple lens, about a millimetre in size, which he could hold quite close to his eye. His method of making it was most ingenious. He had a lamp and some glass; he melted a blob of glass at the end of a thin glass thread and threw the blobs around. In time he collected a great many little glass spheres and he examined them one at a time until he found a good one. There was no going to the trouble of grinding them or any nonsense of that kind! And they were very good lenses. With them he discovered bacteria and spermatozoa. On his death he gave all his instruments – there were forty of them – to the Royal Society. Unfortunately, at that time the Royal Society was as short of money as usual and could not afford to pay the salary of a curator, so, by the time they came round to it, there was little left to curate! For a consideration, a foreign visitor could pick up this or that instrument out of the collection – and nearly all had gone. There is only one of Leeuwenhoek's

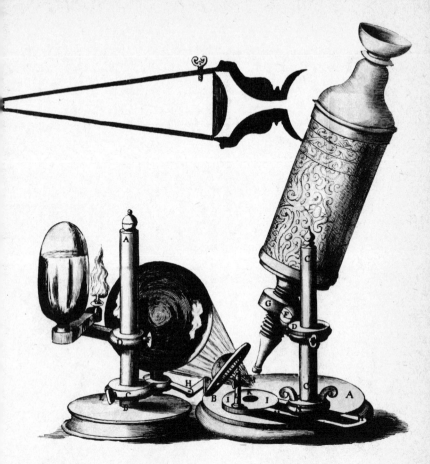

72 Hooke's compound microscope

microscopes left, the others must be somewhere and perhaps one day they will turn up.

Fig. 73 shows us Scheiner's helioscope. Scheiner, who was one of the first of the great tradition of Jesuit astronomers, studied the sun by means of a projection telescope, which makes an image of the sun on a screen, and he discovered sun spots. This caused a great deal of friction between himself and Galileo who did not believe that the spots were where Scheiner said he saw them. Yet sun spot observation has gone on from

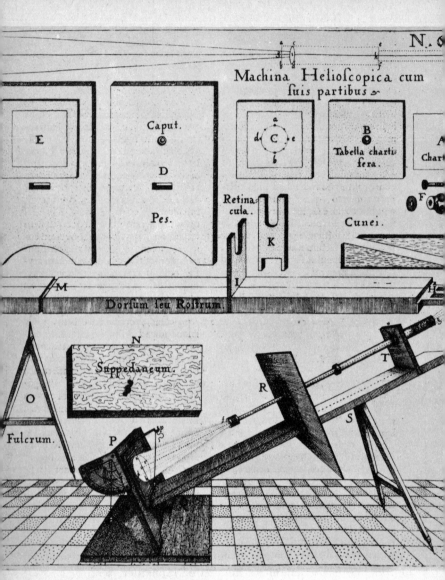

73 Scheiner's helioscope, 1630

that day forward. You can see in this illustration a type of fairly simple real telescope, not a Galilean telescope but with another eyepiece, a convex lens enabling a real image to be thrown.

The problem of the early astronomers was that, when they wanted to do accurate work with a telescope, they were never able to get a good enough image. The images were all coloured, although there were some very good and conscientious lens grinders, including Spinoza, the philosopher, who maintained himself by grinding lenses. But, however well they ground a simple lens, they were unable to eliminate those coloured fringes – and this is what set Newton thinking.

The calculus

After this background to the main point of this chapter, I come to what Newton actually did. I can do no more than roughly sketch this out as it would take more than a whole series of lectures to tell you all the different things that Newton did. To begin with, there is his work in pure mathematics, which I will not go further into here, but in which his greatest achievement was the binomial theorem. From the binomial theorem he got the idea of the calculus. For the variation of a quantity, say velocity, he used his own symbols, which were still used in England but nowhere else up to 1850, of putting a dot, $\dot{x}$, for the first derivative and two dots, $\ddot{x}$, for the second derivative. A slightly younger contemporary of his, Leibniz, hit on the idea of using $\frac{dy}{dx}$, yet, out of loyalty to Newton, it was not until 1845 that in Cambridge you were allowed to use dx and dy. Babbage, or some other wit of the nineteenth century, said that he had rescued Cambridge from the dot-age of Newton and installed the de-ism of Leibniz. The introduction of the calculus made Newton a master at once because, for a long time, he was the only person who could differentiate, and it was so much his own property, so to speak, that when he wrote his great book, the *Principia*, he did not use differential notation as he thought no one would understand it. So he used Greek geometry to prove all the theorems in the *Principia*.

The next thing he did is illustrated in Fig. 74. This shows him as an experimental scientist of quite considerable power.

You can see here Newton's original telescope, which is still fortunately to be seen. It was quite small and it was never much good, but that was due to unfortunate technical error. The important thing was that, in principle, he had solved the problem of how to get rid of the coloured fringes. A mirror has no prismatic or chromatic aberration, therefore if you use a mirror you can obtain a clear image – but the problem then was how to get a mirror. Newton could grind a mirror – the concave mirror, mounted at the lower end, as well as a small plane mirror, opposite the eyepiece – but he could not prevent it tarnishing and, of course, it lost its figure because every time he cleaned it up, it had to be reground. It was so much of a nuisance that it was never much use and there is no record of any observation made with it. But it was a fine idea and it is the ancestor of all the great telescopes of today.

This achievement drew attention to him. As I have said, he was at Cambridge and although he had been awarded a minor Fellowship at his College, Trinity, he was never given a senior Fellowship because there was something very seriously wrong with him. He had incorrect ideas about the Trinity and such a person obviously could not be a senior Fellow of Trinity. For the same reason he could not even be the Principal of King's, and he had to content himself with becoming Master of the Mint, for which theological orthodoxy was not absolutely required. He went up to London with his little telescope and he was received with considerable acclaim in the now well-established Royal Society. He then proceeded to quarrel with various people there – or they quarrelled with him – on the subject of light. I will deal later with his achievements in the field of light. Here, I will discuss his principal achievement which is, of course, the theory of gravitation.

I have given you the descriptive side and I now want to come to the dynamical side. I should have mentioned that Galileo did do something else besides showing the moons of Jupiter. He tried to deal – actually, rather ineffectually – with the physical objections to the idea of a rotating earth. The obvious physical objection was that if the earth rotated, why was there not a terrific wind blowing all round, blowing in the opposite direction to that in which the earth was rotating, from west to east. Such a wind would destroy everything and, therefore, the earth could not be rotating. If you believed that the air reached out indefinitely and filled the whole universe – and obviously there was no reason why you should not believe it at that time – you can see that this was bound to happen. The idea that the atmosphere was limited only came later, with the discovery of the vacuum and the barometer, after Galileo's death. So that people thought they had quite good reason for physical objections. However, Galileo produced arguments against it and, on the whole, even though they were not cast iron arguments, by Newton's time the Copernican picture of the world – the world with the sun in the centre – was more or less accepted. However, it was not in any way explained.

Inertia

Another problem which worried people in the Middle Ages was that of inertia. It was always maintained that in order to get a planet to go round it had to be pushed. As there was no obvious person pushing it, it must be an angel. There was an angel attached to every planet keeping it on its course. Well, if you have an angel, you do not need to worry about timing and such things. The angel has to look after the planet like the driver of a car. The older method was rather mechanical – if you are dealing with the concept of a set of crystalline spheres, you have to have gears, so that you only need one chap outside turning a handle and all the spheres turn round automatically at the right speed. But when you destroy, as Copernicus did, the concept of the crystalline spheres, you have to go back to the angel to get them to go round in their courses. The idea that something went on of itself, especially the idea of something going round and round without being pushed all the time, was inconceivable.

We had to face exactly the same trouble round about the early 1900s, when no one could understand how an electron could get round its orbit, and it was a terrible business to explain it without radiation. An electron is accelerated in its orbit, which has an acceleration towards the centre and, as everyone now knows, since an accelerating electrical charge will generate electromagnetic waves by Maxwell's equation, it must then slow itself down by radiation and fall into the nucleus. To get hold of that idea, all the rules had to be broken – which was done first by Planck and then by Bohr. But the earlier objection to Galileo was quite a serious one: the difficulty of keeping the planets going.

Newton tackled this problem and, although we cannot verify the fact about him and the apple, we now have to take what he said, even in his old age, more seriously. Newton had got, essentially from studying Galileo's work on the pendulum, the idea of centrifugal force. It is usually attributed to Huygens in somewhere about 1673. It is now quite certain, although it was not certain until a few years ago, that Newton had already had the idea in 1665. It seems rather odd that though he had this rather important idea, he never considered it worth mentioning to anybody. However, he did write it down. (Fig. 66, p. 190.) There is another sheet in another book which is completely complementary to it and it takes something to make head or tail of it. He works out the proportion of the force between the earth and the sun to the force of gravity – quite a long calculation. He works out the whole theory of gravitation from a circular pendulum and he takes his values, his actual numerical values, straight from Galileo's *Discourses on Two New Sciences* which nobody knew before and which Newton never spoke of at any time. There is no doubt about it, the actual figures are the same – for they are both wrong! Galileo was somewhat casual about the absolute values of the distance of the earth to the moon – though, of course, he could not help it at the time – and of the earth to the sun. So everything came out wrong, although the idea was right. It has often been suggested that this was possibly why Newton gave it up at this stage.

The study of the swinging pendulum led to the study of the circular pendulum; the study of the circular pendulum led to the idea of the centrifugal force and that, in turn, led to an idea of a gravity holding planets in, this while they swing round perfectly freely. The gravity, in the case of the pendulum, is simply the component of the weight of the bob towards the centre. Once Newton had grasped that, he could work it all out. He was a mathematician but one wonders why he had to work to eighteen figures, all of which are written down; since, as at least the second figure is wrong, the others are not terribly significant. Nevertheless, they were all there. From this he came to a very effective conclusion, for instance, that the force of gravity will move a body in 83,677 minutes through a distance of 100,826,500,737,600 braces – *braccia*, the Italian measurement, which is another evidence that he took it straight out of Galileo's book. Anyway, as I have said, there it was, he had it and he kept it to himself.

But other people were on the track of this too, and the idea that there is a gravitational force which holds the planets in their position, and the idea that things can go on moving, are all ideas which were floating around, although we now know that Newton had them before most people. Huygens had arrived at them as well as Borelli, a very ingenious Italian physiologist, and they would certainly have spoken out but they would not have been anything like as convincing.

The dynamics of the solar system

The object of all this exercise was, of course, to understand the dynamics of the solar system so as to approach the problem in a different way. There are three stages in approaching a phenomena like the solar system. One is simply to note it, to note when the eclipses occur, to note when the stars rise and set, and from that you calculate tables by sheer arithmetic. It is like various aspects of crystallography: in one of them you do not have to know anything about the model – it is actually wrong to have a model – you just know that there are certain recurrences, and if you study the recurrences you can predict on the basis of these recurrences. And this is what the Babylonians did in astronomy. You can then go a little further, as the Greeks did, and make a kinematical model. Instead of an occurrence just happening, they could predict that at 4 o'clock on the evening of 17 September, the star so-and-so will appear on the horizon: 'we have seen it before and we shall see it again'. 'This star', they said, 'is going round, moving this way or that' and they drew a picture of it going round. They did not know why it went round but, they pointed out, it simplified the picture to say that the star is going round rather than just to say that it will arrive at a certain time.

Now, if you have got beyond that and you have the kinematics described as well, as had Kepler, you can still make good tables. But it would be much nicer to know the dynamics of the thing: *why* it moves. Because if you know why it moves you can then work out more accurately what is going to happen.

At least, this was the idea, and there was a strong reason behind it. By Newton's time, no one seriously believed in astrology. Previously, the central importance of the celestial problem was to understand the workings of the divine will in the universe and, although you were not likely to know that, nevertheless it remained the central problem. But now it had a practical – a very much lower – use, to find the longitude, and Newton was very much occupied with this. As I mentioned in an earlier lecture, Galileo had the task of finding the longitude by using the clock and the satellites of Jupiter. But Newton wanted to find the laws of motion, from which he could make good tables, although when he was doing this work, he was not interested precisely in making good tables, he just wanted to get at the principles.

At this stage it would be best to sketch in the ideas, in 1666, on the doctrine of attractions. Borelli, the Italian physiologist I was telling you about earlier, introduced the important idea of the movement of planets which implied the existence of the need to balance the centrifugal force by some other force which he characterised as the force of gravity. This conclusion had already been reached by Newton by that time, in 1665, and both are derived from Galileo. To account for an elliptical orbit, with the planet moving faster as it nears the sun, they argued that the force of gravity must increase to balance the increased centrifugal force. The force of gravity is therefore some function of the power of the distance. The question now became, what function? Hooke, who had already suspected that gravity diminished with the distance, tried to confirm it by looking, though in vain, for the variation in weight in a body on the ground, in a mine-shaft, and at the top of a steeple. He did not find what he sought: his measurements were not sufficiently accurate. Descartes offered another physical explanation of gravity. He pictured the whole solar system as a kind of whirlpool or vortex and considered that, as everything goes to the centre of a whirlpool, so things were sucked in, as it were, but if they had sufficient spin they could keep out. And that was a physical idea that Newton toyed with himself, though later he devoted a whole book of the *Principia* to proving that it was all nonsense, that the laws of motion of a whirlpool, which could be worked out, were not the same as the laws of motion of a planet. But it was a very seductive idea. A whole number of scientists had the idea that the square of the period was proportional to the cube of the radius and it followed, therefore, that the gravitational pull or centripetal force to balance the centrifugal force must depend on the radius divided by its cube, that is, on the inverse square of the radius. This reasoning comes straight out of Kepler's Third Law. The funny thing is that Kepler never saw this. In fact, it is only necessary to know Kepler's Third Law to have the law of gravitation. And many people saw it, not only Borelli but Halley, Hooke and Wren – this by 1679.

The Principia

At that point there were a certain number of little arguments going on and another person suddenly appeared, much younger than any of the others, Edmund Halley, an enthusiastic astronomer who had gone right out to St Helena to catalogue the stars of the southern celestial hemisphere and came back as a great supporter of Newton. In 1684, he sought geometrical proof of the inverse square law of gravitation from Hooke, Wren and finally Newton, who supplied it. There was Newton, who had been sitting in Cambridge, doing what he considered to be much more important work with prisms, with electricity, and other aspects of physics, until he was finally goaded and bullied by Halley into revealing his discoveries. He just sat down and wrote this greatest of books, the *Principia*, the bible of science as a whole and particularly the bible of physics. Unfortunately, now, not unlike the case with the other bible, very few people read it, and nobody, as far as I know, reads it right through. I have not done so myself. I have read quite large chunks of it but I have not read the *Principia* from beginning to end. However, the beginning of it is really good enough and I will read a piece from the Preface, which gives the general idea of it:

Since the Ancients esteem the science of mechanics as the greatest interest in the investigation of natural things, and the moderns rejecting substantial forms of occult qualities have endeavoured to subject the phenomena of nature to the laws of mathematics, I have in this treatise cultivated mathematics as far as it relates to philosophy.

The *Principia* in its full title is 'the mathematical principles of natural philosophy'.

The Ancients considered mechanics in a two-fold respect; as rational, which proceeds accurately by demonstration, and practical. To practical mechanics all the manual arts belong, from which mechanics got its name. But as artificers do not work with perfect accuracy, it comes to pass that mechanics is so distinguished from geometry that what is perfectly accurate is called geometrical; what is less so, is called mechanical. However, the errors are not in the art, but in the artificers. He that works with less accuracy is an imperfect mechanic; and if any could work with perfect accuracy, he would be the most perfect mechanic of all, for the description of right lines and circles, upon which geometry is founded, belong to mechanics.

This is important when you come to consider later the Einstein theory of relativity. 'Geometry does not teach us to draw these lines, but requires them to be drawn, for it requires that the learner should first be taught to describe these accurately before he enters upon geometry, then it shows how by these operations problems may be solved.' Now this, of course, is what might be called an introduction to the axiomatic method.

But I consider philosophy rather than arts and write not concerning manual but natural powers, and consider chiefly those things which relate to gravity, levity, elastic force, the resistance of fluids, and the like forces, whether attractive or impulsive, and therefore I offer this work as the mathematical principles of philosophy, for the whole burden of philosophy seems to consist of this – [and this ought to be written in letters of gold somewhere] from the phenomena of motions to investigate the forces of nature, and then from these forces to demonstrate the other phenomena. . . .

That is, as it were, the programme of physics, very fairly enunciated at the beginning. We would not entirely agree with it now because, since Einstein, we do not like to talk about forces; although we still spend all our time talking about forces but with a mental proviso that they are not real forces, it is only a manner of speaking. In fact, a great deal of objection to Newton was just because he did talk about forces. There were, even then, people like Leibniz who had the more philosophical view.

Newton then gives an explication, as he says, of the system

of the world. By the system of the world, of course, he still means the solar system. He starts off, bang at it, so to speak, and we get these axiomatic rules laid down in the form of definitions. It begins with the definition of something which cannot be defined! 'The quantity of matter is the measure of the same, arising from its density and bulk conjointly.' You see, he believed you could measure things, you could measure bulk. So we now return to this question of what is the primitive measure. Bulk, then, can be measured, but Newton believed that density is an intrinsic property of matter and that therefore it is a perfectly basic thing to measure the quantity of matter – what we call mass. 'The quantity of motion is the measure of the same, arising from the velocity and quantity of matter conjointly.' He does not quite define how you measure velocity. 'The *vis insita*, [this is a new idea] or inate force of matter [what we call inertia] is a power of resisting, by which every body, as much as in it lies, continues in its present state, whether it be at rest, or moving uniformly forwards in a right line.' He then goes on to define force: 'An impressed force is an action exerted upon a body, in order to change its state, either of rest, or of uniform motion in a right line.' Now we have force defined dynamically. He then goes straight on to consider centripetal forces: 'A centripetal force is that by which bodies are drawn or impelled, or in any way tend, towards a point as to a centre.' Those are the only basic definitions; he then goes on to detailed definitions of centripetal forces, etc.: 'I do not define time, space, place and motion, as being well known to all.' Well, as you know, we have had a great deal of trouble about these terms, and you must realise that it is necessary to go back to the very beginning or even further back than that, to define time, space, place, and such concepts. They are not sufficiently well known to all of us as they were to Newton. He then goes on with his axioms or laws of motion:

I. Every body continues in its state of rest, or uniform motion in a right line, unless it is compelled to change that state by forces impressed upon it.

That is really the same thing as the definition but put in the form of an axiom.

II. The change of motion is proportional to the motive force impressed. . . .

III. To every action there is always opposed an equal reaction. . . .

These are the celebrated three Newtonian laws.

He then relates kinematics to geometry by compounding forces as one would compound displacements: 'A body, acted on by two forces simultaneously, will describe the diagonal of the parallelogram in the same time as it would describe the sides by those forces separately.' That is, he measured forces by the amount of motion they would convey to a body. From that he proceeded to work out the orbits, for instance, this one: 'The areas which revolving bodies describe by radii drawn to an immovable centre of force do lie in the same immovable planes, and are proportional to the times in which they are described.' That generalises Kepler's Second Law, because Kepler's Second Law, unlike his Third Law or, for that matter, the First Law, does not depend on the nature of the force or on what that force depends. It would not matter whether the force was $\frac{1}{r}$ or was $\frac{1}{r^3}$ it would still leave the Second Law unaltered because all that the Second Law means is that the moment of momentum is conserved.

Newton then continues at great length, I do not expect you will read it all. It must be said that it was a book which not only explained the theory but also provided all the figures from which planetary and other orbits could be worked out. In one proposition here it is possible to show how to calculate the orbit of any celestial body from three observations. From the tables produced by means of them could be found the motion of the moon – which is the most difficult of all and, in fact, he was defeated on that one: he was defeated because it was really too complicated to be of any use. You see, if you have good moon tables, you can find the position of the moon in the sky very easily, the moon is crossing stars all the time and it is possible to be extremely accurate; but you need good moon tables and Newton never managed to get these. So, in the end, it was a clockmaker, and not Newton, who solved the problem of longitude. To do this, you can either use some celestial phenomenon to tell you what the absolute time is or, in the intermediate stage, you can use a chronometer, a very accurate chronometer. More than a hundred years after Newton's death John Harrison made several chronometers and in 1764 won a prize of £20,000 that the Admiralty had offered in 1714 for any practical method of finding longitude. That is, the prize was awarded to him but it proved extremely difficult to obtain: he had to go to law to get it paid to him.

Harrison's chronometer was the practical solution until, of course, radio and other aids came along, and no one bothers to look at the sky any more now except to see the odd spaceship or rocket. The whole of the problem of the movements of bodies under gravity was solved and, in the process, all the basic principles of physics. In *Principia*, you get the conception of mass, the conception of variable motion, the differential equations of motion, studies of vibration; the whole of classical physics is contained in this book, and this is only one of Newton's great books. The other, about which I will talk in my next chapter, was in some ways more important to the future of physics. I refer to Newton's *Opticks*.

9

The Nature of Light and Colour

In my last chapter, I spoke about the contribution of Newton to the solution of the essential problem which had been posed by the Ancients, the problem of the movement of the planets in the heavens, in the process of which Galileo and Newton had laid the foundations of dynamics. In this lecture, I propose to deal with Newton's other work, in which he made an almost equally vital contribution, his work on optics. Here again, he was taking up a problem left over by the Ancients, but neither they nor the medieval people had got very far with it.

Optics, so far as the Greeks were concerned, was limited to the study of shadows and mirrors. They had progressed as far as the idea of the mirror, the concave mirror, which formed an image and which could be used as a burning glass. The medieval people had got a little further, to the idea of the lens and, in fact, they had 'sold' the lens, so to speak, as spectacles. From that came the next crucial step of the Dutch, who developed the telescope. The rest of the story of optics is very closely associated with the further development of the telescope and later, to a much lesser extent, with that of the microscope. Curiously enough, the modern achromate microscope is a mid-nineteenth century invention.

I will just remind you of Fig. 71 (p. 200) showing the first use of the telescope in the Greenwich Observatory, where most of the elementary devices of the telescope were to be seen. In Fig. 74 (p. 206) you will see Newton's own contribution – a little reflecting telescope. But Newton was not satisfied with his reflecting telescope. He wanted to investigate how to improve the lens telescope and, in particular, how to eliminate in the images the coloured fringes that he saw around objects in the telescope. For that he started a series of experimental studies which were among the first – not the first, which were Galileo's

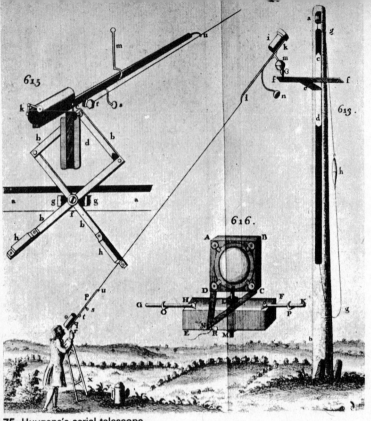

75 Huygens's aerial telescope

– on the practical breakdown of the phenomena of the telescope, of lenses, into their basic elements. This included the study of the properties of the prism. Now, Newton's examination of the prism was really carrying on work that had been done, also in the Middle Ages, not on the prism, but on the result of the prism, that is, the rainbow. The rainbow had always interested people. Obviously, it could be given a spiritual interpretation, as in the story of Noah's flood, but nevertheless all the opticians of the past had worked on it and they had progressed quite far even in the Middle Ages. They had got as far as the idea that the colours of the rainbow were produced by the sun's rays going through the actual water drops in the sky. They found the angle of the relation of the rainbow to the direction of the particular elevation of the sun and they had found that the red rays were bent less than the purple ones. By Newton's time, the elementary laws of refraction – Snell's work – had shown the relation between the sines of the angles of incidence and refraction, but that was as far as the study of light had got.

The phenomena of the prism

Newton took up the matter and started a series of investigations which he called 'to examine the celebrated phenomena of the prism'. The reason they were celebrated was because people, once they could get good glass – and that was a great difficulty in earlier times, they could not get good or clear glass – found that it looked pretty and they had already, in the seventeenth century, started making chandeliers and other things which showed colours. But the question had never really been studied. Newton began his investigations by making a small hole in the shutter in his rooms at Cambridge, letting in the light, letting it strike a prism and then be refracted on to the opposite wall.

Fig. 75 demonstrates the practical result of the difficulties of the telescope. You could get over the difficulties by, so to speak, walking round them and making the telescope with very, very weak lenses and a very long focus. But the difficulties were enormous, as, for instance, those of using that kind of telescope if there were any kind of wind. There was a string, you see, connecting the eyepiece with the objective and how the image could be made to stay still long enough for anything to be observed is anyone's guess.

Fig. 76 returns to Newton letting the sun through the little hole in his shutter and it is producing the spectrum on the

76 The spectrum of light as demonstrated by Newton; from the *Philosophical Transactions of the Royal Society*, 1672

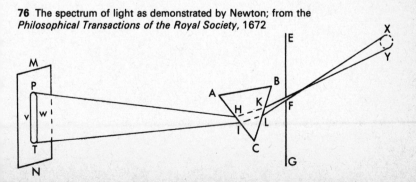

piece of white paper on the wall of his room – and it was there that Newton first described the colours. It is interesting that he never really tried to get a very good spectrum, due probably to the fact that, like many great people, what he failed to observe was in many ways more important than what he did observe. Newton's trouble, even when he was young, was that he did not have very good sight. He never saw anything very clearly and he usually employed somebody else to describe what could be seen. The other person did not describe anything odd about this spectrum and we had to wait for Frauenhofer, the better part of 200 years later, before anyone saw that there were black lines across the spectrum. Therefore, the modern part of the theory of light, that part of the theory that led to the quantum theory, was completely hidden from Newton, though for all immediate practical purposes it did not matter.

Newton drew a rather startling conclusion from the first experiment, that it was the light itself which contained rays, as he saw them, of different refrangibility. Now, this concept of rays was about the oldest of concepts. It was the light beam which, in the old days, was thought to proceed in the inverse direction, from the observer of the thing observed. But the idea that light was a kind of projectile, whichever way it was shot, was the basis of Newton's thought. As we see, there was a period in between when there was a certain amount of argument as to whether light was waves or particles. What Newton was dealing with was the movement of what we call photons. He now saw that the process could be inverted and this was one of his first ways of doing it (Fig. 77). He passed the white

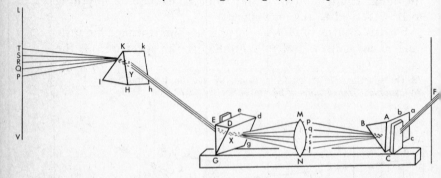

77 Newton's experiment to demonstrate the recombination of spectral colours to form white light

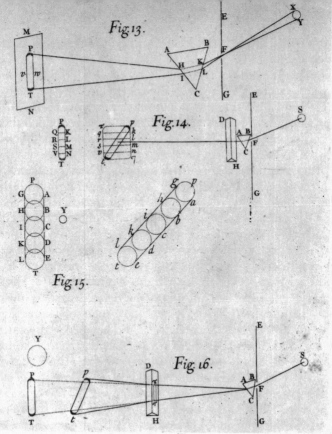

78 Illustrations of Newton's experiments on the refraction of light, taken from his *Opticks*, London, 1704

light coming from the right through a prism which cast a spectrum on the lens which, in turn, focused the coloured rays on a second prism arranged so that it would neutralise the effect of the first prism and send a beam of white light to the third prism which again formed a spectrum. Thus he showed that white light could be turned into a spectrum, be recombined into white light and then turned into a spectrum again.

Fig. 78 illustrates the process of showing that nothing had happened to the light and that the rays which were originally more refracted could be shown to be more refracted in a second prism. The vertical spectra on the left show what it would be like without the second prism. Newton put the second prism in at right angles and was able to give a picture of the spectrum which is obliquely reflected. You see there his full understanding of the analysis of this according to the colours, the more or less refracted rays being the blue or the red.

Now, he did not publish the results of all this work for many years, although he did the experiments in the early 1670s. He published them as individual papers of the Royal Society in 1675 and he then wrote his book *Opticks* of which the second edition appeared as late as 1718. Although this book is, in a way, a much smaller book than the *Principia*, it was for physics almost equally important, because in it Newton did not confine himself to this problem of the refraction of light. He tried, first of all, to resolve it into rules. He proposed sets of definitions and axioms which you all learn: the angles of reflection and refraction lie in the same plane; if the refracted ray is returned directly back to the point of incidence, it should be refracted into the line previously described by the incident ray – that is the complete reversibility of light. But here he says: ' Refraction out of the rarer Medium into the denser is made towards the Perpendicular; that is, so that the Angle of Refraction is less than the Angle of Incidence.' Then he goes on to state Snell's law. But here there is a rather subtle point. Atmospheric refraction had been known almost since the time of the Greeks. They had already noted that if you try to make accurate observations of stars, you must never study them near the horizon because the times do not fit in there, the star always takes a bit longer to set and rises a bit sooner than expected. Now the Greeks had already hit on the idea that the atmosphere had something to do with it. This was particularly important, actually, because a great deal of dating was based on the time at which a star rose. When, for instance, Sirius, the Dog-star, rose at the same time as the sun, that established the period of the year and a few days later the Nile flood occurred. It was one of these practical utilisations and therefore tables of refractions had been constructed for the atmosphere.

What Newton did, however, was a very interesting example of how research might be undertaken for a certain purpose and, in fact, completely fail to achieve the objective. Newton considered that the amount of refraction was proportional to the amount of what we now call dispersion, that is, the difference of refraction between the red and the blue rays. From this he concluded that it would never be possible to make an optical instrument that did not show the colours and that, therefore, the problem he had set himself was intrinsically insoluble by lenses. He studied a great number of substances and found this to be the case – only roughly, as you see – and even made some rather interesting physical deductions from it. For instance, he noticed the very high refractive index of diamond and the high refractive index of oils, and this made him think that diamond must consist of oily material, in other words, that it must be composed largely of carbon. But at that time no one had ever done an experiment with the burning glass on diamond, nor could they have understood it, in any case, because they had no knowledge of elementary chemistry.

Newton's failure even in that respect was also very important. Curiously enough, it was never put right by a scientist. It was put right by optical people, by Dollond in the first place, after Newton's death. He found that, although Newton had said it was impossible, if you put two pieces of glass together with different refracting power, they could compensate and this led to the achromatic objective in about 1758 and made an enormous difference to astronomy. Newton's interest, however, went much further than that. He was very interested to find out about other aspects of the behaviour of light. But he was not the only one in the field. I seem to be harping on the failures of Newton, but one of them, most interesting from the point of view of later theory, was that Newton all the while thought of light as particles and, therefore, he thought that light going from a thinner medium to a denser medium was being speeded up by attraction as it got into the denser medium. At that time, actually during Newton's lifetime, the velocity of light had been established. It is a queer story – one of those discoveries based on a series of accidents.

The velocity of light

In 1671, the French astronomer, Picard, had, for historical reasons, wanted to visit the laboratory of Tycho Brahe on the island of Hveen. He arrived there and found that the local tenants had destroyed the whole observatory and there was nothing to be found there. In wandering around looking for it, he talked to a young peasant, Römer, who seemed to be very interested in the subject and Picard took him back to France.

Römer started his own observations and found that Galileo's theory based on his observations on the rotations of the moons of Jupiter was incorrect. It showed anomalies; at some times of the year the moons of Jupiter were moving faster and at other times they were going slower – they were out of turn with each other. This seemed to be a function, strange though it seemed at the time, of the distance of Jupiter from the earth. Römer then hit on what seemed a fantastic theory, like the relativity or quantum theory, that light had a speed, it took time to move. The thought was not quite as upsetting as it might have been because everyone knew that sound took time to move and many people had tried the elementary experiment of taking the shutter off a dark lantern, telling the chap at the other end to open his dark lantern when he saw the flash. What you measured that way, of course, was the time of reaction of the observer. But that experiment works very well with cannons. You shoot off a cannon and the other man shoots his off when he hears the sound and in that way they were able to measure the velocity of sound.

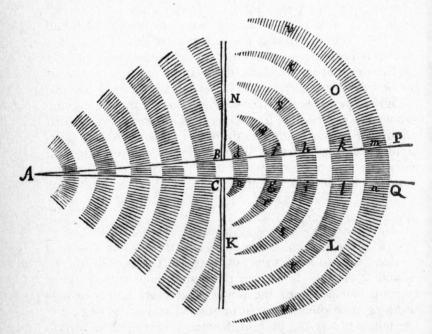

79 Newton's diagram of the particles of light; from *Principia*

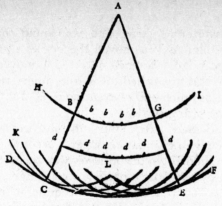

80 Huygens's illustration of the spherical wavelets of light; from *Traité de la Lumière*, Paris, 1690

Christian Huygens

It seemed there was a definite velocity of light and, therefore, there could be different velocities of light in different media. It was not until much later, however, that it was found that Newton's assumption that light travelled faster in the denser medium was wrong. Now the alternative view of light was to imagine that light was like a sound, that light was a vibration. This was the view taken by Newton's great contemporary Huygens, the Dutch physicist and astronomer who, oddly enough, as Holland was at war with France over most of the period, worked in the French Academy. Fig. 79 illustrates Newton's idea of the way in which light went through a hole and produced a new set of what looks very much like waves. But Newton did not think of them as waves at all, he thought of them as particles in different states, as you will see in a moment.

Fig. 80 shows Huygens's picture with which you are all familiar. That is, if there is a wave of light or, for that matter, a wave in water or a wave of sound, each part of the wave front may be deemed to send out a set of circular waves, the envelope of which becomes the next situation of the wave, from which you can deduce a number of things about interference, diffraction and so forth. The first test of the two theories came as a result of the properties of a very curious substance, Iceland spar, found by a mineralogist of that time, Bartholin. It was a great wonder; here was a magic crystal, you put it on a table over some print, and you found that the print appeared double. As you moved the crystal round, the two images moved relative to each other. One moved more than the other and this was called the extraordinary ray; the other did not move as much and obeyed the rules of diffraction and was called the ordinary ray.

Newton had made great efforts to solve this problem and failed completely to give any reasonable account of it. Indeed, obviously he could not in the ordinary way because it could not be explained by any simple particle view of light, although he got as far as saying something that was quite interesting and is now much more significant. He said that the ray of light may have sides to it, that is, it might not be symmetrical around the ray. One side might be one way and the other side be different, which is really what we call a polarised beam. But apart from that he could not deduce anything at all accurate about Iceland spar. Huygens, however, could.

Huygens explained the normal laws of refraction in a diagram which is very interesting to us nowadays because it was the one used by the Braggs and others for X-rays (Fig. 81). In this, a wave coming down to a medium of different wave velocity, in this case the denser medium being of lower wave velocity, the wave front gets bent in the way shown. The contradiction is absolute. In Newton's theory the light wave has to move faster down below, in Huygens's it moves slower. Fig. 82, which perhaps should have been placed first, is Huygens's construction for the more elementary phenomenon, reflection.

81 Huygens's construction for the refraction of light

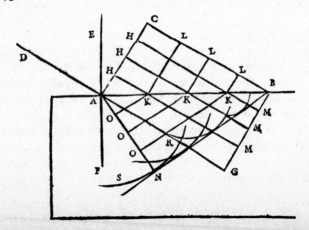

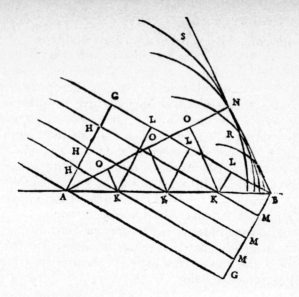

82 Huygens's construction for the reflection of light

What Huygens suspected was that in a peculiar medium it might be that the speed of the movement of the wave was dependent on the direction in which it moved and the little fronts, so to speak, could become ellipses and not circles, and the wave, therefore, inclined. I am thinking here of the military term – right turn is right incline. In this Huygens's construction (Fig. 83), the wave comes down as a wave which is parallel to the surface, that is, normal incidence, and it is still at normal incidence when it emerges. But an individual ray comes down and moves, as shown, to the left of the figure so that when it hits the other side it comes out at a different place but moving in the same direction. That is the characteristic

83 Huygens's construction showing the behaviour of light in Iceland spar

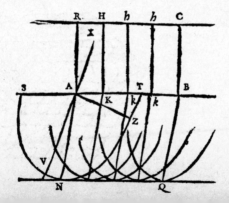

of the movement of a wave in a medium of this kind.

In Fig. 84 Huygens's real genius is illustrated. Iceland spar occurs in these rhombic forms which are like cubes, but are not quite cubes because three of the angles shown here are not right-angles, they are obtuse angles. Huygens imagined that a cubic crystal was made out of spherical atoms, and that a crystal of this sort was made out of atoms that were oblate spheroids which could fit together and make a rhomb and this rhomb would necessarily have different optical qualities in different directions. In other words, he gave a complete and correct description. However, it had to wait for Sir Lawrence Bragg actually to calculate the differences. In modern parlance these oblate spheroids are not atoms but ions, the carbonate ion, CO_3, which is oblate, and here you can actually calculate from the elementary refractivity of the oxygen atoms the change in refractive index.

Newton's rings

I now come back again to Newton. The colours we have been considering are not the only colours that Newton looked for. He came across something else that proved to be of even greater importance. Everyone had seen the colours of, let us say, oil on water, or the colours to be seen in mother-of-pearl, opal and other substances, where there are thin layers that seem to produce colour. How does this occur? Newton made the observation that you all know very well, that of so-called Newton's rings. If you put a nearly flat lens on top of a glass plate and look at it from the top and from the bottom, you find a set of rings (Fig. 85). He noted very accurately what the

84 Huygens's description of rhombic forms in Iceland spar

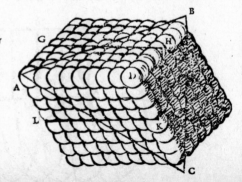

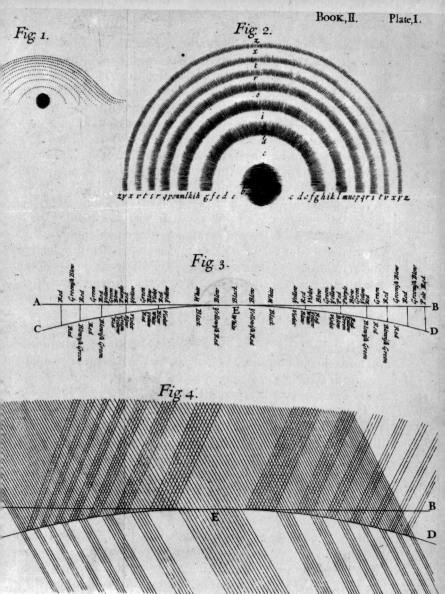

85 Newton's rings

colours are – white in the centre, if you look *through*; but when looked *at*, it is black in the centre, then goes blue, then white, which is the first order; then yellow, red, violet and blue again; these are the various orders. Newton sorted them out, but it is the theory that is so interesting. How, he asked, can a particle

229

give rise to different colours according to the different thicknesses it goes through between two refracting substances? Newton's deduction was a very strange one. He said that it was possible that the particle as it goes along is in different fits. He meant by 'fit' that it is a fit of easy reflection or easy refraction. The particle, he argued, changes its mind as it goes along; it goes some way in a 'fit of easy reflection' and goes back again; if it has further to go, it will get into a 'fit of easy refraction' and it will go through. If it goes further still it will get into a 'fit of easy reflection' and go back again. If you look at it in monochromatic light, which Newton, to his disadvantage, did not have, you will see that then you would get a series of what we call monochromatic fringes which are due to interference. But Newton's concept of these 'fits of reflection and refraction' are really what we would call the phases of the vibration. The vibration is in different phases, which was what Newton was measuring in his statement 'the distance between fits is a function of the refrangibility of the rays'. That is to say, the blue ray has many fits per inch, twice as many, let us say, as had the red ray. In that way Newton laid the foundation of the idea of the frequency of a wave, just as Huygens had laid down the idea of the geometry of the refraction of a wave in a non-homogeneous or polarising medium.

There was a great deal of trouble over Newton's optical work, trouble which continued long after his death, because people did not like the idea that white light was not a pure notion. Everybody could see that white light was purer than red or blue light and this idea that white light was a mixture of all the colours offended people. Incidentally, in the history of science, one of the last of the people it offended was Goethe, who had a poetic objection to white light being composed of colours. He believed that colours arise when white light is mixed with darkness. Thus, when white light passes through haze, yellow or even red light is transmitted. The blue of the sky arises when the darkness of space is observed through haze illuminated by daylight.

Newton was fundamentally what we would call an objective scientist, he was concerned with what light did and not what it was observed to do. Goethe looked at it quite subjectively and, whereas Newton laid the basis for what might be called objective optics – the kind of optics that can be put on a photographic plate, or a photon counter from which can be obtained the right answers – Goethe was really laying the foundation of subjective optics, that is, the way the eye sees things. In the same way there are two acoustics: the actual sound that is produced by the musical instrument, the sound track, and what is heard from the musical instrument through the instrumentation of the ear, which picks out various harmonics. And, so, the eye itself and the various nervous mechanisms which go with it do not give a true representation of light.

The interesting thing that Newton did after this, which turned out to be of even greater importance than the optical part of the work, was to write down his opinion on the nature of matter which I will discuss later in this chapter.

Diffraction

Fig. 86 illustrates still other colours produced by light going through narrow holes. This is the beginning of the study of

86 The diffraction of light; from Newton's *Opticks*

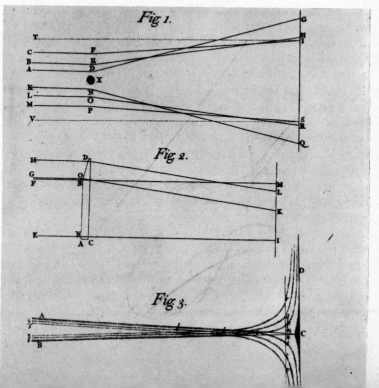

the diffraction of light. It shows a pin and the light going by it and being, as it were, bent away from it – according to Newton's views – in a really physical sense. This means that anything going near an edge, partly bounces off it: you can see that the ray that comes nearest bounces furthest and from this you get the idea of the effect of a slit. If light is put through a wedge-shaped slit, it shows greater and greater scattering as the slit gets narrower and narrower. This was first observed by Grimaldi. It was one of those experiments that people must have done for thousands of years and never paid any attention to it. If you look at a sharp bright light through a feather, a piece of cloth, or something similar, you can see all these diffractions. Newton established the essential features of it, which were that diffraction depends on the narrowness of the aperture and on the quality of the light; that is, the blue light is diffracted more than the red light. If Newton had used a grating, this would have given him the whole theory of diffraction.

Newton only got as far as this in his study of light and, in a sense unfortunately, it was so good and so well laid out in his *Opticks*, that it continued to be used for more than a hundred years and no one really did anything better or anything different from Newton with the exception of the practical application of the making of an achromatic lens that I mentioned earlier. In fact, when the theories of light were reversed again, due to the work of Young and Fresnel in the 1820s, Young had very serious difficulties in getting his views accepted because they were against the canon of Newton's *Opticks* and he was attacked very strongly in the *Edinburgh Review* by one of the founders of this college of ours, Lord Brougham – you will find his name in the entrance hall of Birkbeck – for daring to attack the authority of Newton. Young, in fact, countered fairly well by quoting extracts from Newton which indicated that Newton was not absolutely, one hundred per cent, in favour of the corpuscular theory of light but inclined somewhat – by the suggestions of 'fits' and the idea that light had sides and periodicity – to a wave theory of light.

However, as often happens in the history of science, people's doubts and second thoughts get eroded away and only the simplest Newtonian doctrine got across. Newton, himself, was well aware that he had not come to anything like the end of the subject and he includes in his book a series of queries. I do not think that any very important person has ever published queries since, but they were very common in the seventeenth century and then they faded out on the grounds that there was quite enough published on what people knew, without wasting a great deal of paper publishing what they did not know! What Newton did not know turned out to be almost more important than what he did know.

I should like to pick out for you some of Newton's queries which show how his mind was working and where it led him. He starts with light, asking: 'Do not Bodies act upon Light at a distance, and by their action bend its Rays; and is not this action strongest at the least distance?' That was the question of diffraction. 'Do not the Rays which differ in Refrangibility differ also in Flexibility. . . . Are not the Rays of Light in passing by the edges and sides of Bodies bent several times backwards and forwards, with a motion like that of an Eel?' He then goes on: 'Do not Bodies and Light act mutually upon one another; that is to say, Bodies upon Light in emitting, reflecting, refracting and inflecting it, and Light upon Bodies for heating them, and putting their parts into a vibrating motion wherein heat consists?' Now, these are the types of problems which are worrying theoretical physicists even today. 'Do not black Bodies conceive heat more easily from Light than those of other Colours do, by reason that the Light falling on them is not reflected outwards, but enters the Bodies, and is often reflected and refracted within them until it be stifled and lost?' He goes on from here to chemical speculations: 'Is not the strength and vigour of the action between Light and sulphureous Bodies . . . one reason why sulphureous Bodies take fire more readily, and burn more vehemently than other Bodies do?' A not altogether happy query there!

He then talks about various forms of emission of light. Now, in those days, very naturally, everyone was concerned about the nature of fire, and this was not to be solved in the first place by physics but by chemistry.

Very curious types of people occupied themselves with this problem; one person, Voltaire, who wrote a treatise on fire, carefully heated up bodies and weighed them and was able to show that their weight did not differ when they were hot or when they were cold and that heat was not a material substance. The experiments were probably completely wrong because there were no precautions taken about the effect of the air on the weighing, but as he used large cannon shot it probably did not make so much difference. Voltaire asked: 'Is not a flame vapour, fume, or exhalation, heated red hot, then so hot as to shine?' He went on at some length about these things and, there is no doubt he did a great deal of experimental work on them.

The other side of Newton's work, to which I will refer later, is expressed in his query: 'Do not great Bodies conserve their heat the longest, their parts heating one another, and may not great dense and fix'd Bodies, when heated beyond a certain degree, emit Light so copiously, as by the Emission and Reaction of its Light, and the Reflexions and Refractions of its Rays within its Pores to grow still hotter, till it comes to a certain period of heat, such as that of the Sun?' A curious idea, this. Bodies reach a kind of critical point at which they automatically heat themselves by emitting light. He then goes on to discuss physiological matters about the effects of light on the retina of the eye and so to physiological optics: 'May not the harmony and discord of Colours arise from the proportions of the Vibrations propagated through the Fibres of the optick Nerves into the Brain, as the harmony and discord of Sounds arise from the proportions of the Vibrations of the Air?' This is something that many people followed up but, unfortunately, there is nothing to it. He did things like pressing his finger on his eye in the dark and found that he got a generation of light. He goes on talking about waves and then to the problem of how light gets through the medium through which it is travelling:

Does not the Refraction of Light proceed from the different density of this Aethereal Medium in different places, the Light receding always from the denser parts of the Medium? And is not the density thereof greater in free and open Spaces void of Air and other grosser Bodies, than within the Pores of Water, Glass, Crystal, Gems, and other compact Bodies?

You see, he had to invert the whole idea of things. He would have liked light to move fastest in open things, so he argues that 'light pays attention not to the density of matter but to the density of ether, and ether is much less dense inside matter than in free space. That is why light moves faster when it gets into a solid medium.' Of course, this question was settled in the nineteenth century by the actual measurement of the velocity of light going through water and the velocity of light in a dense medium was also worked out.

Newton goes on to speculate about the nature of the action of muscle; in fact, there is very little in physics or biophysics which does not appear somewhere or other in Newton's books. He goes on to the unfortunate discussion of the nature of Iceland spar. Later he discusses the nature of cohesion of bodies, and he says:

Now that part of the resisting Power of any Medium which arises from the Tenacity, Friction, or Attrition of the Parts of the Medium, may be diminish'd by dividing the Matter into smaller Parts. . . . But that part of the Resistance which arises from the *Vis inertiae* is proportional to the Density of the Matter, and cannot be diminish'd by dividing the Matter into smaller parts, or by any other means than by decreasing the Density of the Medium.

Incidentally, this is a very curious phenomenon which is true and was only really worked out fairly recently in relation to very high velocity movements. You see, if the velocity of a movement is not very high, the properties, the Young's modulus or something else, of the medium really determines what it does. But if the velocity is very high only the density comes in. For instance, if you have an explosion, a high velocity shock wave, and you have some quite loose powder, you can turn it – a shock wave travelling about 4,000 feet per second – through a right angle by a piece of thin tissue paper or by powder of the same density.

Newton concludes with much discussion of chemistry but, at the end, comes the question of the definition of the atom:

Have not the small Particles of Bodies certain Powers, Virtues or Forces, by which they act at a distance? . . . All these things being consider'd, it seems probable to me, that God in the Beginning form'd Matter in solid, massy, hard, impenetrable, movable Particles, of such Sizes and Figures, and with such other Properties, and in such Proportion to Space, as most conduced to the End for which he form'd them; and that these primitive Particles being Solids are incomparably harder than any porous Bodies compounded of them; even so very hard, as never to wear or break in pieces. . . . But should they wear or break in pieces the Nature of Things depending on them, would be changed. Water and Earth, composed of old worn Particles and Fragments of Particles, would not be of the same Nature and Texture now, with Water and Earth composed of entire Particles in the Beginning. And therefore, that Nature may be lasting, the Changes of corporeal Things are to be placed only in the various Separations and new Associations and Motions of the permanent Particles: compound Bodies being apt to break, not in the midst of solid Particles, but where those Particles are laid together, and only touch in a few Points.

These are almost word for word definitions of atoms given by Gassendi, and effectively the same as those given by Democritus 2,000 years before.

It seems to me further that these Particles have not only a *Vis inertiæ* accompanied with such passive Laws of Motion as naturally result from that Force, but also that they move by certain active Principles, such as that of Gravity, and that which causes Fermentation, and the Cohesion of Bodies. These Principles I consider not as occult Qualities, supposed to result from the specifick Forms of Things, but as general Laws of Nature, by which the Things themselves are form'd, their Truth appearing to us by Phenomena, though their Causes be not yet discover'd. For these are the manifest Qualities, and their Causes only are occult. And the *Aristotelians* give the Name of occult qualities not to manifest qualities but to such Qualities as they supposed to lie hid in Bodies, and to be unknown Causes of manifest Effects: Such as would be the Cause of Gravity, and of magnetick and electrick Attractions, and of Fermentations, if we should suppose that these Forces or Actions arose from Qualities unknown to us, and uncapable of being discovered and made manifest. Such occult Qualities put a stop to the Improvement of natural Philosophy, and therefore of late Years have been rejected. To tell us that every Species of Things is endow'd with an occult specifick Quality by which it acts and produces manifest Effects, is to tell us nothing: But to derive two or three general Principles of Motion from phænomena, and afterwards to tell us how the Properties and Actions of all corporeal Things follow from those manifest Principles, would be a very great step in Philosophy, though the Causes of those Principles were not yet discover'd: And therefore, I scruple not to propose the Principle of Motion above-mention'd, they being of very general Extent, and leave their Causes to be found out.

Now by the help of these Principles, all material Things seem to have been composed of the hard and solid Particles above-mention'd, variously associated in the first Creation by the Counsel of an intelligent Agent. For it became him who created them to set them in order. And if he did so, it's unphilosophical to seek for any other Origin of the World, or to pretend that it might arise out of a Chaos by the mere Laws of Nature; though being once form'd, it may continue by those Laws for many Ages. For while Comets move in very excentrick Orbs in all manner of Positions, blind Fate could never have made all the Planets move in one and the same way in Orbs concentrick, some inconsiderable Irregularities excepted, which may have arisen from the mutual Action of Comets and Planets . . . Such a wonderful Uniformity of the Planetary System must be allowed the Effect of Choice. And so must the Uniformity in the Bodies of Animals, they having generally a right and left side shaped alike. . . .

Well, I can only quote a little more as space is limited:

As in Mathematicks, so in Natural Philosophy, the Investigation of difficult Things by the Method of Analysis, ought ever to precede the Method of Composition [what we call synthesis]. This Analysis consists in making Experiments and Observations, and in drawing general Conclusions from them by Induction, and admitting of no Objections against the Conclusions, but such as are taken from Experiments, or other certain Truths. For Hypotheses are not to be regarded in experimental Philosophy. And although the arguing from Experiments and Observations by Induction be no Demonstration of general Conclusions; yet it is the best way of arguing which the Nature of Things admits of, and may be looked upon as so much the stronger, by how much the Induction is more general. And if no Exception occur from phænomena, the Conclusion may be pronounced generally. But if at any time afterwards any Exception shall occur from Experiments, it may then begin to be pronounced with such Exceptions as occur. By this way of Analysis we may proceed from Compounds to Ingredients, from Motions to the Forces producing them; and in general, from Effects to their Causes, and from particular Causes to more general ones. . . .

The book ends with a sort of philosophical caution. I will give you the last words as these are the last words Newton ever published.

And if natural Philosophy in all its Parts, by pursuing this Method, shall at length be perfected, the Bounds of Moral Philosophy will also be enlarged. For so far as we can know by natural Philosophy what is the first Cause, what Power he has over us, and what Benefits we receive from him, so far our Duty towards him, as well as that towards one another, will appear to us by the Light of Nature. And no doubt, if the Worship of false Gods had not blinded the Heathen, their Moral Philosophy would have gone farther than the four Cardinal Virtues; and instead of teaching the Transmigration of Souls, and the worship of the Sun and Moon, and dead Heroes, they would have taught us to worship to our true Author and Benefactor....

Newton's philosophy

Now, this general philosophy of Newton is extremely interesting in two directions. First of all on the scientific side: you see, what it really comes to is that Newton had formed a new picture of the universe as a universe of law but a universe completely without the concept of change or evolution. In fact, his atoms have to be the same because Nature must always be conformable to itself, it cannot change. The second thing is that everything must have been set in the first place. The interesting thing is that Newton put in a few exceptions in relation to comets and odd movements of the planets. Laplace, about a hundred years later in his *Celestial Mechanics*, showed that you did not need any exceptions, that if you had formulated the gravitational conditions correctly, everything fell into place. In fact, it was Napoleon who said to him, 'Well, I do not see anything about God in your thesis.' Laplace replied, 'I have no need for that hypothesis'.

The fact is that Newton had an essentially conservative view, and an essentially unhistoric view. For instance, he thought it a very remarkable fact that all the planets go round the same way in a plane, and this could only have been done by some definite pre-design. Some fifty years later, Kant, and a hundred years later, Laplace, showed that a disc of gas and dust would take that elliptical shape, and of course as a particle in an orbit has to go round the same way, everything else would have to go round the same way. Newton had a similar concept about the elements. He considered the elements, the atoms, as permanent, and in one sense he was quite right: if the atoms change, everything made out of them changes. If you decompose a nucleus of oxygen into its four constituent helium nuclei, it is no longer oxygen. But the concept that the solar system might have a history, and that the animals (which he brings in and invokes at this point Aristotelian-like in that the right side and the left side of an animal should be more or less the same) are products of evolution, had to wait until Newton's ideas were led one step further by Darwin 150 years later.

10

The Development of the Steam Engine and the Theory of Heat

In this chapter I propose to deal with the original object of physics, which was correcting astronomical observations and producing proper predictions from them. To a very large extent the things I talked about in my last chapter – the optical investigations and experiments of Newton – were all really done with a view to the improvement of telescopes for astronomical observations and it was in that rather exacting period of science that all the very essential mathematical foundations, both of optics and of mechanics, were laid. Curiously enough, however, it was in quite another field, which I have not touched on so far, that the real break-through was made which was to lead to developments of science of more practical utility.

So far, all the things I have spoken of were of practical utility to seamen and navigators only but to no one else. Now came the first break-through which joined the Scientific Revolution, which I have been discussing, to the Industrial Revolution that was soon to come. It arose out of a technique which is connected essentially with mining, namely, water lifting. The real curse of mining is not the getting of the metal or the coal out of the mine, it is getting the water out of the mine. There is always anything up to twenty times as much water to be lifted as there is anything else; and this is bound to be so because once a hole is made in the ground all the water around flows into it. You will remember the illustrations of those very ingenious pumps, simple pumps with buckets, more complicated pumps, suction pumps, force pumps – they were all developed for mining and they all required an enormous amount of power. It was very fortunate that there was water-power for this.

The functioning of pumps was essential in many parts of the world; particularly, for instance, in Holland where most of the best parts of the country were under the mean sea level and had to be pumped out by windmills. But the wind for windmills is not generally reliable. Thus, one of the most essential things required was an alternative source of power and then to get the pumps to work properly. Now, the practical people had known for years that an ordinary kind of village pump with a handle – a suction pump – would not suck water up more than thirty feet. To go deeper it was necessary to put one pump underneath another, the one pumping into the sump and the next one pumping out of that. There had either to be a man on each pump or the pumps had to be linked together at the top to make them all work together. This was a serious limitation to the possibility of pumping, quite apart from the amount of energy involved. Around 1630, Galileo, who was the prime mover in many of these things, looked into this problem with his two new sciences and came to an entirely erroneous conclusion. He knew that if you tried to put down, let us say, a forty-foot pipe and put a suction pump at the top, you could pump – and it was quite a labour to pump – but nothing would come out of it. He assumed, therefore, that a water column was produced in the pump and that when it got beyond a certain length it broke. The tensile strength that water could hold, he thought, was limited, the water column inevitably broke and, very naturally, the water just fell right down to the bottom again.

The vacuum pump

Fig. 87 shows what was in effect the first vacuum pump. EFGH was the piston and the space above it was filled with water. There was a conical valve to which a bucket was attached into which sand was poured to act as a weight. The weight at which the bucket started to come down was then measured. By this means Galileo attempted to show what happened to a water column, but what he was really measuring was atmospheric pressure – apart from all the leaks around his piston. He never got beyond that and the pump did not seem to work very well. But he had a student called Torricelli who, in 1640, had the brilliant idea of using mercury instead of water. The great advantage of doing this was that it could be done in a glass tube. You see, no one knew what was happening

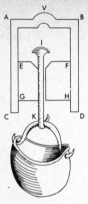

87 Galileo's apparatus for attempting to measure the resistance to the formation of a vacuum. EH was a wooden piston in the hollow cylinder AD. The space above the piston was filled with water and the bucket filled with sand until it started to descend

inside the barrel of the pump as it could not be seen. They did not know whether the water was there when they tried to lift it or if it had fallen to the bottom. Mercury, on the other hand, could be seen and Torricelli found that the mercury would go up to the corresponding gravitational height – that is, thirty inches instead of thirty feet, the ratio of the densities being roughly 13:1 – and above the mercury was a gap. Now, this gap gave rise to an enormous amount of controversy – it apparently had nothing in it! If it had nothing in it and there was no way of getting anything in it, it must be what the Ancients had declared to be an impossibility: it must be a vacuum. Nature, we were told, abhorred the vacuum, but Torricelli showed that the abhorrence was a relative one and nature stopped abhorring it at thirty inches of mercury or thirty feet of water! So it was not long before the elementary idea dawned on them that they were not really measuring the horror of a vacuum, they were measuring the pressure of an atmosphere which was pressing the water – or the mercury, as the case may be – into the tube and that thirty feet of water or thirty inches of mercury would more or less balance the pressure of the atmosphere.

That was the first barometer and, of course, its truth was checked: if you went up higher from the bottom of the atmosphere, then it should register not thirty inches but less. This was, in fact, Pascal's experiment, one of his long-range experiments because he did not actually carry it out himself but got his brother-in-law, Périer, to carry it out for him on the summit of the Puy de Dôme in Central France. Périer did it very carefully: he left a barometer at the bottom of the mountain, carried one up and found that the difference in atmospheric pressure corresponded to the weight of that height of air. From then on, of course, the barometer became the basic instrument for meteorology because people had noticed that it changed from day to day or even from hour to hour and, apart from the details of different methods of reading, the barometer has not altered very much to the present day. But it did mean an enormous change in attitude because now it had to be admitted that a vacuum was possible.

About this time, in the first half of the seventeenth century, a very long war – in fact, the Thirty Years War which lasted until 1648 – was being waged in which various engineers had the chance to show their mettle and to a certain extent profit from it. One of them was the Mayor of Magdeburg, which town, incidentally, was entirely destroyed in that war and had to be rebuilt. The Mayor, Otto von Guericke, supported the Protestant side and he became Quartermaster General to the Swedish King, Gustavus Adolphus. With Tycho Brahe, he was one of the first scientists to have enough money to carry out reasonable scientific experiments. His first idea was that to get a vacuum was quite simple: all that was required was to take a barrel, fill it with water, pump the water out and there is your vacuum. Fig. 88 shows an ordinary syringe pump with the washer turned the other way – it can be done with a bicycle pump – and the plunger is just pulled out. This led to rather

88 Von Guericke's attempt to pump water out of a wooden barrel; from *Experimenta Nova*, Amsterdam, 1672

89 Von Guericke's improved experiment: pumping air out of a metal sphere; from *Experimenta Nova*

unexpected results. First of all, it was very hard work pulling out the water and by the time, as they reckoned, they had half the water out, the barrel started sizzling and it was quite clear that air was getting in through the cracks. They failed to get a vacuum and the barrel was full of air when they had finished. So they thought they would prevent this by putting the barrel inside another barrel, which did diminish the amount of air but now they had filled it with water instead of air. They then realised that barrels would not do, so they started with a thin copper sphere – and then they had a shock! When they started pumping that out, the whole thing collapsed; it dented in. However, by making a thicker sphere they got it to work. Fig. 89 shows the same very crude way of pumping, but now they had a sphere of bronze made in two parts, and they also had the possibility of detaching the sphere from the pump by having a stop-cock on it which they could open or close at will.

Having got these spheres, they began to do demonstrations (Fig. 90). They tried to measure the atmospheric pressure by the elementary method of hanging the sphere up and putting weights on it to try and tear it apart. But a better demonstration,

90 One of Von Guericke's experiments with Magdeburg hemispheres; from *Experimenta Nova*

91 (overleaf) The famous demonstration by Von Guericke of the power of the vacuum. The two teams of horses were unable to pull the hemispheres asunder so long as the vacuum was maintained; from *Experimenta Nova*

Fig. I.

A

B

H

Fig. III.

Fig. II.

Fig. I.

or a more successful one, carried out before the Emperor, consisted of having eight horses a side (Fig. 91) with the sphere in the middle, and this really showed what a powerful thing the vacuum was. Fra Angelica immediately had the idea that this could be used for transmitting power. If you had pipes of this kind, he argued, you could make a pump or a signal or something similar work at a distance by having a vacuum tube – the principle of the vacuum brake. But with the techniques of the time it could not be done because there were too many leaks. However, the idea created, as it was intended to create, a great deal of interest in scientific circles and various people in other places began to study the problem.

One of these people was the Hon. Robert Boyle, who is referred to as the father of chemistry. He was a son of the first Earl of Cork and had money enough, at least, to employ a very good laboratory assistant, Hooke. It is strongly suspected that the actual experiments were all carried out by Hooke. Boyle meditated about them and wrote down the answers but it is significant that after Hooke left Boyle, Boyle never did any more experiments.

Fig. 92 illustrates a very fundamental experiment by Von Guericke of a type much derided by Charles II. There are two things of note in this picture: first, the small sphere shown is balanced by a weight and when the stop-cock is turned, the sphere suddenly goes down. In other words, they are weighing the air, because the sphere was empty before and now it is full of air. Charles II, with Nell Gwynn and other people, had great fun deriding the philosophers of the Royal Society when they occupied themselves with such idle pursuits as weighing the air. The other part of the picture shows a water barometer, going up, as can be seen, to the fifth storey and, although it is very sensitive, it is not very reliable because of the water vapour. Enlarged on the left you can see the index, a little man floating on a cork on top of the water column.

Fig. 93, again, shows some of the barometers of the time including another example of a water barometer which was made by Boyle.

Boyle decided to make his own pump and his first effort is

92 Two of Von Guericke's experiments on pressure of the atmosphere: weighing air (front centre); water barometer (rear centre and right; left front)

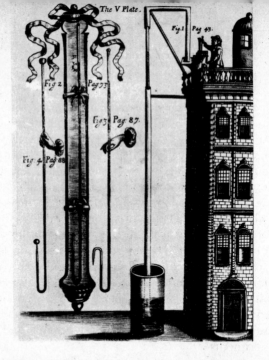

93 Boyle's experiments on 'the spring of the air'; from *A continuation of New Experiments, Physio-Mechanical touching the Spring and Weight of the Air*, Oxford, 1669

a very good example of development (Fig. 94). He wanted to see what happened inside a vacuum and he made the receiver of glass, necessarily very thick glass. The cylinder can be seen dismantled on the right. In order to save himself the actual force required for pulling on a string, he had a rack and pinion device to pull out the air. It was very hard work as the piston kept springing back, but the interesting thing is that he devised a way of introducing things into the receiver, an opening at the top with an air-tight stopper. This was the basis of a whole scheme of science that touches on the chemical side. It was here that Boyle discovered that things would not burn in a vacuum, that animals would not live in a vacuum, that air was necessary for both combustion and respiration, and this is the basis of the whole of modern chemistry, which is centred on the chemistry of respiration and burning. Boyle, in that sense, was definitely the father of chemistry. This development is called the pneumatic revolution in chemistry, and you will

94 Boyle's air pump using a glass receiver to view the inside of a vacuum; from *New Experiments* . . .

fig: 13

fig: 12

fig: 1

A

B   C
D   E
K

K

fig: 2.ª

fig: 6.º

fig: 7

fig: 9

fig: 16.ª

fig: 10.

fig: 5.ª

fig: 14.ª

fig: 15;

fig: 8.

know from the study of ordinary elementary chemistry how the constitution of the air, the constitution of water, etc. were found out essentially by pneumatic experiments.

For the moment, however, we are concerned with physics. The next improvement was to separate the receiver altogether from the actual cylinder. Fig. 95 shows this separation. The experimental animal, probably a rat, is in the receiver, but we now have added to the same elementary rack and pinion, a water trough to prevent the back leakage of air. This is still very inefficient. Boyle was not satisfied and he continued work on it.

95 Boyle's second air pump; from *New Experiments* . . .

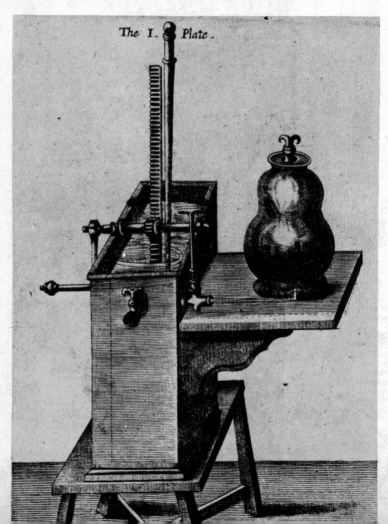

The I. Plate.

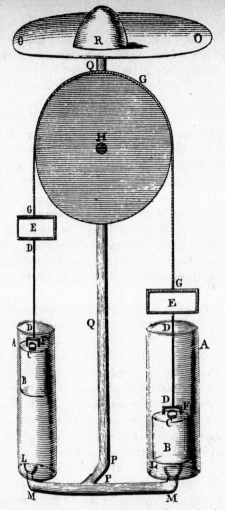

96 Boyle's third air pump, which was operated by depressing iron stirrups EE, alternately with the feet. Air was pumped out of the receiver R on the plate O. From the *Works of R. Boyle,* 1772

Fig. 96 shows the fundamental difficulty of having to pull against the atmospheric pressure, a difficulty which is partially overcome by balancing one cylinder against the other. This, incidentally, was about as far as the vacuum pump developed until the twentieth century. When I went to the Royal Institution to work in 1923, we had a pump just like that, which we used to get our back vacuum: you had to work it up and down as shown in the illustration. After a few minutes you could get a reasonable vacuum and then a diffusion pump could be put on the top. Boyle was very experienced in these kinds of pumps, but then he went further.

The spring of the air

One of the most noticeable phenomena in working a pump of this kind is what Boyle called, very graphically, 'the spring of air'. If you try to move a piston in or out, it jumps back at you but, depending on whether the air is compressed or sucked out it jumps back one way or the other, just as if it were attached to a spring. It was necessary to measure this spring of air. This measurement Boyle did by the classical experiments which you all repeat in the most elementary physics courses, with a U-tube pouring in mercury at one side and, and this is the interesting thing, making careful note of the height of the mercury in each tube, the volume of the air and so forth, and trying to see whether these figures mean anything. From that you get Boyle's Law, that is, the product of pressure and volume is the same. As Boyle always worked at the same temperature, that is as far as it got.

Boyle's Law is the first physical law in science: all the other laws are essentially mathematical laws or, at the most, mechanical laws. But this was a law of physics because the explanation of Boyle's Law was not at all obvious, in fact it did not come until much later. The first real explanation was given by Maxwell about three-quarters of the way through the nineteenth century. In the old days, they often thought of the particles of air as having actual little springs on them, and when measuring the spring of air they thought they were measuring the springs that held the particles apart in the air. Boyle, at any rate, made the measurement and this was an enormous advance as it led to the possibility of making use of the compressibility of gases.

Now, even before Boyle had measured the spring of the air, people had thought about making use of it. I have mentioned the need for power, particularly for power in pumping, and one of the expressions used all through the seventeenth century was 'raising water by means of fire'. Everyone knew that fire had great power and they wanted to use it for raising water. There are various ways of doing this but one obvious way would be somehow to make a vacuum – once the idea of the vacuum is established – and then to get rid of the vacuum; that would draw up the water and then, of course, the water has to be pushed down or allowed to run down again – that is one way of pumping.

Denis Papin

Christian Huygens, mentioned in the last chapter, had as his technical assistant a man called Denis Papin from Blois. Papin was very concerned with just this problem of the pump, but he started by examining in a very scientific way the process of boiling. Through this he immediately hit on a very good idea which, unfortunately, was only fully used two hundred years after his death, and this is what is called Papin's digester, the original pressure cooker. He constructed a strong iron vessel with a lid (Fig. 97) which was held down by brackets and screws and what is so interesting – his own invention – a safety valve. He found that usually, when the whole vessel was put on fire, it blew up, but with the safety valve on, it stayed put. This safety valve was employed in all his subsequent devices.

97 Papin's Digester, the original pressure cooker; from Robert Routledge, *A Popular History of Science,* 1881

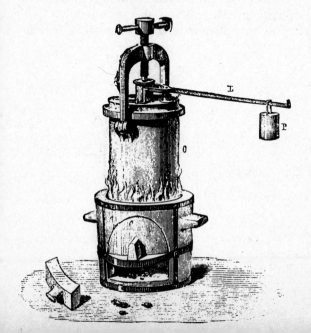

Fig. 98 is a picture of a very interesting piece of apparatus. This is really the first internal combustion engine, of Huygens. It is obviously not a working drawing but just a sketch. There is a piston D which is attached to a weight at the top over a pulley. There is the cylinder AB. Down at the bottom there is a little capsule, C, which contains a small piece of gunpowder, and the gunpowder is ignited by means of a hot rod which is introduced. The gunpowder blows up, up goes the piston, then the air and exhaust gases are let out, and the piston comes down again and pulls the weight up on the other side. Essentially the same engine was being used about 1840; it was the first oil engine and was ancestor of all the internal combustion engines of today. Though gunpowder is not a very suitable material either from the point of view of cost or otherwise, yet it avoids the difficulty of having to have an air intake. If you have gunpowder, all you need is the solid fuel at the bottom, which supplies the gases necessary to push out the piston.

Fig. 99 shows Papin's first effort at a steam engine. As you see it contains a boiler, a steampipe and a free piston. A small quantity of water is placed at the bottom of the cylinder A, which is heated from the bottom. The steam produced pushes up the piston, B. A latch, E, catches in a notch in the piston-rod H and holds it until released. The heat is removed, the steam condenses and creates a partial vacuum, E is released and the atmospheric pressure drives the piston down. The whole idea was how to get something which worked without

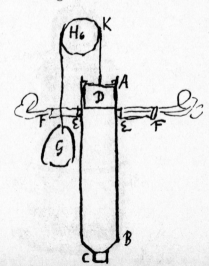

98 Huygens's sketch of the first internal combustion engine; from *Oeuvres Complètes,* 1897

99 Papin's steam engine apparatus

any elaborate moving parts. Papin was not successful, unfortunately, because he could not get enough money. In 1708,
he wrote a very pathetic letter to the Royal Society asking for
fifteen pounds to perfect his steam engine which would be of
great use to everyone, but the Royal Society refused to give
him anything on the grounds that unless he demonstrated that
the engine really worked he did not deserve to get any money
– and he never did. Well, that was how scientific development
was encouraged in the eighteenth century. But other people
who were a bit shrewder did manage to get some money. The
idea of the engine – what was called a philosophical engine –
was a scientific one, but the scientific people were not the right
people to develop it on two grounds. First of all, they did not
understand the techniques sufficiently well and, secondly,

they did not understand the economic aspects – what it would pay to do.

Captain Savery

The next person to tackle this problem was a man called Captain Savery; the same principle was used but now the piston was not needed, the engine worked directly without a piston. The steam just pushed the water down, the water was forced into a receiver and when the receiver cooled, the water was sucked up; then followed alternate sucking and pumping – a process which is used to this day in a device which is called a pulsometer pump. It is lowered down a well and pumps the water out of it. It is convenient from some points of view because it has a very simple and rugged construction and very dirty water can be pumped out with it because there are no cylinders or pistons or parts to foul.

In Fig. 100 you can see how it actually worked. You will notice that it has become quite elaborate. It has now become double acting; there is a boiler, which has to be nearly at the bottom of the shaft and, in fact, this was the practical disadvantage of this machine. It would work in a small shaft, but in a 200 to 300 foot mine shaft, all the machinery would have had to be put down at the bottom, which is not the right place for it. The man shown working it alternately directs the steam into one or the other chamber and while the steam is pushing the water up in one chamber, the vacuum is sucking the water into the other.

Savery, himself, was something of a scientist but mainly a considerable inventor. He invented a paddle-boat but he was not the first person to do so. It had been invented during the Wars of Liberation of Holland about a hundred years earlier and actually used in warfare – an armoured paddle-boat worked by men on treadmills inside – it went right through the Spanish lines and it could not be stopped. In the old days, of course, galleys were great sport, in a battle you might cut off all the oars, but a paddle-boat with paddles well-covered could defy that kind of thing. Savery invented one kind and he also invented this pumping machine. He was quite good at dealing with the publicity side and issued a pamphlet called *The Miner's Friend*; the illustration in Fig. 100 is copied from this.

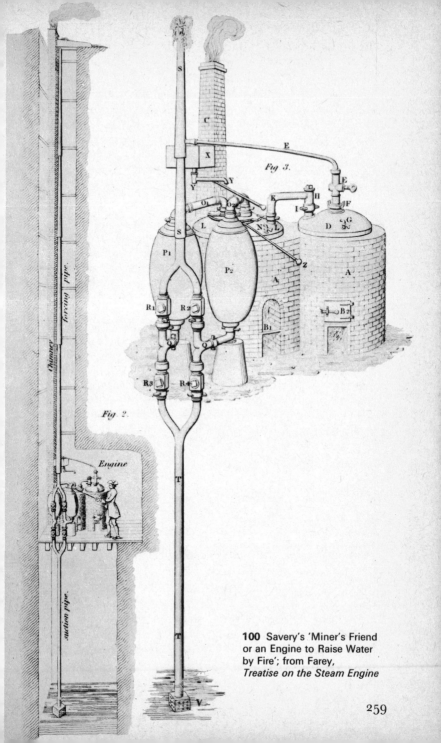

100 Savery's 'Miner's Friend
or an Engine to Raise Water
by Fire'; from Farey,
Treatise on the Steam Engine

In his pamphlet, not only did he describe the engine but he pointed out how useful it would be for different purposes. The following extract is from the preface addressed to 'the Gentlemen Adventurers in the Mines of England'. The mines were essentially royal mines but anyone could 'adventure', that is, put up all the money for the actual sinking of shafts and working the mine and you had to be a gentleman to do this because otherwise you would not have enough money:

I am very sensible a great many among you do as yet look on my invention of raising water by the impellent force of fire a useless sort of project that never can answer my designs or pretensions; and that it is altogether impossible that such an engine as this can be wrought underground and succeed in the raising of water, and draining your mines, so as to deserve any encouragement from you. I am not very fond of lying under the scandal of a bare projector, and therefore present you here with a draught of my machine, and lay before you the uses of it, and leave it to your consideration whether it be worth your while to make use of it or no. . . .

For draining of mines and coal-pits, the use of the engine will sufficiently recommend itself in raising water so easie and cheap, and I do not doubt but that in a few years it will be a means of making our mining trade, which is no small part of the wealth of this kingdome, double if not treble what it now is. And if such vast quantities of lead, tin and coals are now yearly exported, under the difficulties of such an immense charge and pains as the miners, etc., are now at to discharge their water, how much more may be hereafter exported when the charge will be very much lessen'd by the use of this engine every way fitted for the use of mines?

Now this is a very good piece of advertising and it has the unusual feature of being quite true because, in fact, the steam engine did not only treble, it more than quadrupled the output of the mines as the result of its development.

Thomas Newcomen

But it was not to be Savery's steam engine, it was to be the steam engine of an ironmonger from Dartmouth, Newcomen, who, very wisely as Savery had the patent, went into partner-

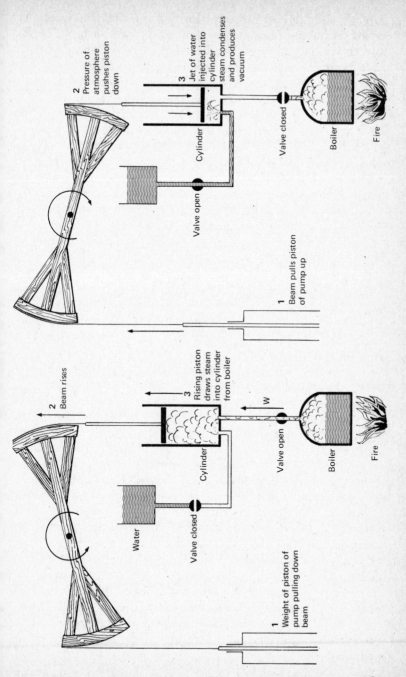

2 Beam rises

2 Pressure of atmosphere pushes piston down

3 Rising piston draws steam into cylinder from boiler

3 Jet of water injected into cylinder steam condenses and produces vacuum

W

Cylinder

Cylinder

Water

Valve closed

Valve open

Valve open

Valve closed

Boiler

Boiler

Fire

Fire

1 Weight of piston of pump pulling down beam

1 Beam pulls piston of pump up

ship with him. He actually built an engine which was essentially based on the old mining type of pump engine: it had a beam and the piston at one end and the other end was pulled down by a crank (Fig. 101). The difference between Savery's and Newcomen's is that the latter is a piston engine, but it is also a vacuum engine – that was why it was called a philosophical engine, because the development of something like this could not have been conceived by a common workman.

Well, no doubt you have all seen these things in the Science Museum in South Kensington where they can be seen working. They were vast machines and had to have a huge shed to house them; and a machine of this kind produced approximately four horse power. It was probably the most expensive capital expenditure per horse power ever, but you have to start somewhere. Its efficiency was incredibly low, probably one or two per cent, but if it was to be used in a coal mine, the cost did not matter; there was always plenty of coal, fuel costing virtually nothing, and with coal came iron. They did not worry about the workmanship very much: the argument was that a piston could be considered a good fit if you could not drop a sixpence between the piston and the cylinder. The way they overcame the bad fit was to cover the piston with a layer of water; true, a certain amount of water entered with every stroke but that helped the condensation and they just kept pouring more water into the top of the piston.

Fig. 102 shows quite a large engine – this is more or less a contemporary drawing. You will notice that the boiler is an ordinary brewer's boiler. Savery had to make things which would withstand reasonable pressures, $\frac{1}{2}$ or $\frac{1}{3}$ atmospheres; if the engine was to be down a sixty-foot pit, he would have had two atmospheres on it. But the Newcomen engine only required steam at atmospheric pressure and that was much safer and easier to do. The piston had to be fairly large and a man was needed to operate the valve. It was an art; you had to know exactly when to turn the valve because the water kept sinking in the boiler and the fire was uneven. The direction in which the beam was moving had to be watched, and the valve turned. The funny thing is that the valve gear on the early engines was much more difficult than on the later ones because the low pressures at which it was working made it very tricky to operate.

102 Savery's engine, incorporating a brewer's boiler

103 Newcomen engine erected near Dudley Castle in 1712; from an engraving by Barney, 1719.

Fig. 103 is a slightly improved variety of the same thing, as you see, in 1712, which is quite early. But this engine has acquired what was necessary and absolutely fundamental to the history of technology, a new, very elaborate valve gear, which I have not the space to describe. The man has disappeared completely from this engine – it is now automatic. There was a man needed to work Savery's engine, but no man is needed here at all. The gear enables the whole engine to work completely automatically at a rate controlled by the rate at which steam can be raised.

We will now consider a little science instead of technology. While all this was going on, we are speaking of 1712, scientific development was not moving very fast. It had moved very fast in Boyle's time but it had declined in the period described in the song of the Vicar of Bray: 'When George and pudding time came in'. The general intellectual level of England fell in the first half of the eighteenth century and rose very much in the latter half. Some of the engines we have been considering were expensive but they paid off, some of them lasted 150 years with the same engine working the whole time. People were quite satisfied with them and there were no improvements. The engines were even used by the relatively rudimentary scientists of the period and that is, in fact, where the new developments arose.

Latent heat: Joseph Black

Science was not taught in England at all, really, but it was taught in Scotland and in the Universities of Edinburgh and Glasgow there were courses in natural philosophy. The Professor of Chemistry at Glasgow University was a medical doctor called Joseph Black who, among other things, laid the basis for the study of the nature of gases. But here we are concerned with his other scientific achievement. On a fairly cold day, the weather suddenly turned a little warmer and all the snow lying outside his place began to melt. He thought to himself – a thought that must have occurred to many people before – why does it not all melt at once? The temperature had risen above zero but all the snow was not melting, it was melting gradually and, he thought, why gradually and not all at once? He then did some experiments with ice and snow and discovered what he called latent heat, that is, heat contained in the water. In order to turn the ice or snow into water, there had to be heat in it which did not cause any change in temperature. Then, very naturally, as the main industry of the time in Scotland, although not very advanced, was the manufacture of whisky, Black made investigations in the whisky distilleries and found that vast quantities of cold water had to be put into the worm of the still. He thought, where does the heat come from? The mash has to be heated and the heat comes back in the worm. Therefore, the heat that is put in remains latent in the steam and turns up again in the condenser.

He was also able to measure this heat. He measured the latent heat of fusion of ice, and that of the condensation of water and he also measured the specific heat of things over a considerable range of temperature. After all, three and a half times more heat is required to boil water away than to bring it to the boil.

He used to lecture on this and he collected a rather curious laboratory assistant – he was not really a laboratory assistant, he was an independent gentleman – called James Watt. James Watt was the son of a carpenter who had prospered and built houses and ships. James Watt, being enterprising, went south to London to become a mathematical instrument maker. He became a very good craftsman, making quadrants, theodolites and other instruments. He then returned to Glasgow and wanted to set himself up but was told that this was quite impossible as he was not a Freeman of the City of Glasgow and only Freemen were entitled to set up shops to sell these things.

James Watt: the separate condenser

However, he was not put off by this. He found that the University was not within the jurisdiction of the City of Glasgow and he managed to persuade the University to give him a little room in the quadrangle for a workshop and in return for that he helped around the place. In the Department of Natural Philosophy there was a model Newcomen engine which was used annually for lecture demonstration, but which had never worked. The students were always told, 'What a wonderful thing is the Newcomen engine, and now we will demonstrate the way it works' – and it never worked! Everyone, including the professor, thought there must be something wrong with its mechanism. It had been sent to London for repair but returned no better than before. In 1764, it was given to Watt to mend and after spending some little time on it, he realised that it would never work because all the heat was being used up in a very strange way. He was puzzled until he learnt from Professor Black of the latter's discovery of latent heat.

So Watt constructed a model condenser – which is all that is left of his original work – and measured its capacity and that led him to the basic improvement. At first sight it looks very much like the earlier one; you have the same arrangement but if you have a large cylinder, a yard across, the thermal losses through the wall are very small. If you have a small cylinder, the losses are very large. Watt overcame this difficulty, even on the large scale, by putting a steam jacket round the cylinder. The essential thing, however, was that, instead of throwing water into the cylinder and cooling the whole thing, walls and all, then having to heat it up on the next stroke, he cooled the steam in a separate condenser. You see, the strokes previously had to be pretty far apart to heat the whole cylinder and cool it down at each stroke. This could not be done with a small model: the essential thing was to do the cooling in a separate condenser. Watt's engine (Fig. 104) was the same old beam engine except that now there was some thermal efficiency in it. It jumped from about two per cent to about six or seven per cent efficiency and this made all the difference. The point here is that the old engines did not get any further than they did because they were so inefficient that they could only be worked where the coal cost nothing, that is, in the coal mines themselves. But by that time there were other mines, the tin mines in Cornwall particularly, with the copper mines there just emerging, and the Newcomen engine could not be used there, it just would not work economically.

Matthew Boulton

Watt, however, had an unfortunate life despite his work. He was later taken up by a great pioneer of the times called Roebuck, the founder of the modern chemical industry, who also helped to found the first iron works in this country, the

104 Boulton's and Watt's rotative beam-engine

Carron Iron Works near Falkirk in Scotland. Roebuck had the right technical ideas but he was not sufficiently business-like, and the Carron Iron Works went bankrupt. Among his assets were Watt and his engine which, incidentally, had never worked. Roebuck was a Birmingham man and he brought it to the attention of his old friend Boulton, who was the greatest industrialist of his time. Boulton owned the largest button factory in Birmingham, but although this factory was suitable for its purpose up to a point when it was mostly a matter of fiddly little bits of metal work, when he got going in a big way, it needed some power and he acquired a millpond and a mill. But Birmingham is not a very hilly district or a very wet one, and the millpond always dried up in the summer time. That was sufficient for working in a small way because men could then be temporarily dismissed during the summer when they could work in the fields bringing in the harvest, and that was the way industry was run until 1870.

Roebuck suggested that Watt came along with his engine so that it could be used to pump the water up from the millpond into the mill race and so work the wheel. This he did and Boulton made a great deal of money out of it. This gave Boulton an idea, his ideas were very important scientifically, or, rather, the ideas of Boulton and Watt between them were. Boulton produced the money but he also produced an enormous amount of push, which was very necessary to get things done because Watt was not a very pushful person, but a rather melancholy Scot. Until he was eighty years old, he said he had never had a day free from anxiety about money.

Boulton's idea was that he would sell something that no one had ever sold before – power. He actually used those words; he wrote to Empress Catherine of Russia saying, 'I am selling what the whole world wants: power'. And this is how he did it. He sent his people down to Cornwall to say: 'We are offering engines on these terms. Our firm, Boulton and Watt, will set up the engines, free, gratis and for nothing, at your mine. We will service them for the first five years and all we are asking in return is one-third of the difference between the cost of coals for our engines and the cost of hay for the horses that would have to do the same amount of work.' Well, the mine-owners thought he was obviously crazy but they accepted his offer.

Now, of course, came the disputed question of how much work a horse could do. First of all it is necessary to determine what is work? Galileo and Newton never discussed 'work'; it is a very odd concept, you see, it is not this or that inertia, it has not a Latin name, it is just work. There were two terms used, one of which has gone out of use and that is 'duty' – the turn of duty in a treadmill or in anything else, was work multiplied by time. Watt measured the amount of work a horse could do by making a horse pull something lifted over a pulley. He conceived the idea of work being the product of force and distance and of power being the rate of doing work. All these, as we think, elementary concepts of physics and thermodynamics, really came entirely from practice, they all came from Manchester or Birmingham people and they were not really recognised in London, as you will see later on. Having done this, Watt also very wisely secured a patent and, by various political pulls of one sort or another, had the patent extended for a fantastic period of forty years, therefore holding up anyone else's development, unless they were prepared to cheat – which, of course, a great many people did.

Meanwhile came the next problem. The mines were all right; the mine-owners complained, of course, after the first ten years, about having to keep on paying Mr Boulton and Mr Watt; it seemed to them they would go on paying for ever. They had quite forgotten what the situation was at the beginning. No one wanted horses now, everyone wanted engines. The trouble with an engine, however, was that though it could pump water, it could not actually make anything go round; and the reason for this was that the action of the cylinder was so irregular. Here, then, was where Watt introduced two developments which transformed the whole situation.

Fig. 104 (p. 268) shows one of the original rotative machines which actually combined the feature of an old beam machine with a rotating part. It is a very curious thing, a sun-and-planet gear. The part with the planet on it goes round and round and turns the wheel at half the speed. We might ask: why could he not use a crank? Well, the reason was, believe it or not, that the crank was patented by somebody else. Watt was so furious that someone had patented the crank that he determined not to use one and not to pay the royalties. In fact, they came round to cranks afterwards; I do not know what happened to the patent, it probably expired. So we now had the first modern engine which was a real rotary engine, one that could be used for the new factories, which did not so much want water pumping as turning of machinery.

There was, however, one element which for the future was to be even more important. You can see what made the rotary

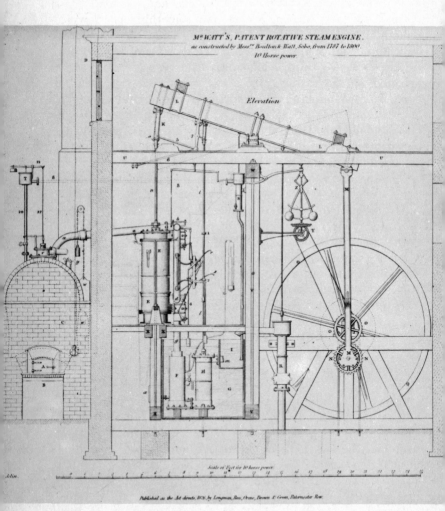

105 Watt's double-acting rotative steam engine of 1784; the sun-and-planet gear may be seen on the right

process possible; the centrifugal governor. If you translate 'governor' into Greek, it becomes *kybernetes*. That is all cybernetics means – governorship, steersmanship – and this governor adjusted the steam inlet to the cylinders according to how rapidly the machinery was going. This is the first example of real feedback. The rest is all valve gear and so forth, Watt's rather ingenious devices to drive a piston straight up and down from a beam instead of hanging it from a beam. All the previous machines worked on tension, they could not be driven, you could just pull one side and then the other. To drive them needed one of these gears.

Fig. 105 shows a steam-engine which was used in a big flour mill; it still has this sun-and-planet gear with the governor and by now it is double-acting, it has the cylinder which admits steam at both ends and it really is quite a remarkably efficient engine. From this time on things never looked back.

The locomotive

So far, all this, apart from the condenser, was done without any idea of the science of heat beyond Black's measurements of latent and specific heat; but it is a rather curious fact that the whole of modern thermodynamics really arises from the steam engine and not *vice versa*. The next stage was introduced by a man called Trevithick, and was contemporary with the developments we have been discussing. Trevithick reversed the whole process; he did not care for condensers; what he wanted was direct action. It occurred to him that if you worked with high pressure steam – all these existing machines were working with low pressure steam – the advantage of a condenser was relatively small and, therefore, you need only use the most elementary process of letting the steam into the cylinder, just pushing the piston out, letting it back and pushing the steam out again. This was of very little use for a mill where you had to economise in coal, but, of course, it led to a far smaller engine. Here were these big machines – Watt's engine rose up to quite twenty horsepower or thereabouts – but they had one fatal disadvantage for ordinary uses, they were so heavy. The next desire was to have a locomotive engine, an engine which could actually move around. For that it was necessary to have an entirely different kind of engine, and it was basically the Trevithick engine which pushed things around fast but inefficiently.

This led to the railways. The rails are not at all new; they go right back to classical times but, of course, they were wooden rails. The rails were usually in mines but as there were mines in Newcastle, which are fairly near the sea, it was very convenient to extend the railway out of the mine and let the trucks run down to the quays or where the ships were loaded up with the coal. Another mine might be on the side of a hill so the trucks could run down: they would have one of these stationary engines at the top which would pull the load up to the top of the hill and then allow it to run down on the other side. These were all railways.

One of the men who operated these engines, a man called Stephenson, had a son called George who thought he could improve on this performance. He was an almost unlettered character – I mean in contrast to the people I have been talking about until now, Watt and the others, who were really scientists of a kind. Watt understood the principles not only of mechanics but of physics, chemistry and much else, whereas Stephenson had learnt nothing at all. He is supposed to have learnt his letters on the shovel while his father was stoking the fire at the engine, but he had ideas; he was a good mechanic and he had a little engine made, a little locomotive engine. There were lots of other people doing it at the time; it was essentially a beam engine to start with which just turned the wheels round. No one knew much about it, and they did not know whether the wheels would slip or whether they would actually go, because no one had ever driven anything along the ground by turning the wheels; they had pulled things with wheels but had not driven them with wheels. Stephenson had his engine running along by the road at about two to four miles an hour and he walked alongside it, chasing the animals off the line and so forth. But even that was not satisfactory because the horses shied at this awful object going along the road, especially every time it shot out a jet of steam, and the local landowners said it just could not be allowed; it must be prohibited altogether.

Then Stephenson did something which was a fundamental development of engines. He turned the exhaust pipes from the cylinders into the funnel in order to prevent the exhaust from scaring the horses and he was enormously surprised to find that his engine, which previously had a speed of about four miles an hour now ran away from him. No one had ever travelled at such a speed before, it was faster than the fastest horses. This meant that the locomotive had really arrived and it led to a terrific locomotive boom. And it led to something else. This was 1828 and a French engineer named Sadi Carnot, who was very learned – quite the opposite of Stephenson – began to wonder how it was done, how it was possible to convert the energy of a fire into the energy of the wheels. He produced a little book on the motive force of fire,* which laid the foundations for the whole of thermodynamics. He went back to the study of the simpler engines, although they were really derived from the study of the locomotive engine, where there was a condenser, and the heat was transferred to the water, and then to the steam and then to the condenser, but not all of it. Some of the heat went from the condenser and came out as motive power. This concept of heat as motion is, of course, one of the oldest, in fact, *the* oldest part of physics because it goes back to the Old Stone Age. In the Old Stone Age that was the way they obtained fire in the first place, by converting work into heat by rubbing sticks together. Everybody knew this phenomenon but when they became more learned they began to question it and the person who put them right was a rather odd character called Benjamin Thompson (Count Rumford) who, incidentally, founded the Royal Institution. He actually measured the amount of heat that you can get by boring cannon and established, but not very accurately, the mechanical equivalent of heat. Carnot also worked out the mechanical equivalent of heat – that is, the first law of thermodynamics – in the little book to which I have just referred. He postulated the second law of thermodynamics, that is, that the efficiency of an engine is proportional to the difference of temperature between the boiler and the condenser, and he gives tables. This book has something I discovered myself, incidentally, but it seems to have escaped everyone else's attention. He gives tables which enable you to work out what the equivalent will be if the temperature of the boiler were

*Réflexions sur la Puissance Motrice du Feu, Paris, 1824.

infinite. This, of course, is the mechanical equivalent of heat, the reverse way. The reason it was important is that the engines were always being improved but no one had any theories; they did not know how far they could be improved. Even the idea of the efficiency of an engine was an idea introduced by Carnot, and from that arose the whole of the modern idea of the theory of heat, the whole of the thermodynamics and essentially the whole of modern physics.

We still have to deal with one other large branch of physics – electricity and magnetism.

11

Electricity and Magnetism

We now have one aspect more to discuss in the formation of classical physics and that is in the field of electricity and magnetism. We have witnessed, as it were, the departure from the old classical physics, the physics of the Greeks, which was concerned very largely with astronomy and to a lesser extent with light and movement. Now we come to the phenomena which, although noticed for a long time, had not been seriously studied or used at all in the older times. They are mostly new phenomena.

The first one I have mentioned already and will not say much more about, that is, the elementary properties of magnetism. Magnetism was probably noticed almost as early as iron, because it is just in those areas where iron is found that the black stone which attracts it, the magnetic stone, was to be found. Later, it was called the lodestone, when its directive property was discovered. No one knows, even now, how the lodestone becomes magnetised. This often happens in science: a subject can be studied in detail, and its principles established, but the actual observation that leads to its study may be beyond analysis. This you will see is also true about electricity and amber.

However, the first important scientific discovery about magnetism was the one I alluded to earlier (p. 126), that is, the directive power of the lodestone. To put it the other way round, it was that the earth was a magnet. This was first realised by the Chinese; from China it passed to Europe, and the first scientific treatise there on the subject was that of Peter the Pilgrim in 1269. In it he explained the ordinary properties of the magnet: that it induced magnetism; that when cut in two it becomes two magnets with two sets of poles – all this was known.

Robert Norman

Yet there was something new to find out, which was not discovered until the end of the sixteenth century, the dip in the compass. The dip was discovered in a very practical way – most discoveries were made in this manner. In the old method of making compasses, the needle was first cut out of iron and balanced on a pin-point, after which it was magnetised by rubbing it with a lodestone. On trying to re-balance it on the pin-point, one end of the needle – usually the north end – was found to dip downwards; but this could be corrected by putting a little weight at the other end. This was what the compass makers had been doing for about 300 years without thinking about it, so to speak, until one of them, Robert Norman, thought he could solve the problem by a radically new experiment, namely, balancing the magnet so that it would move in a vertical plane – and he discovered the dip. This he called the Newe Attractive, meaning a new attractive power of the magnet, to point down as well as to point to the north. He wrote a book about it which really contains most of the matters which were afterwards dealt with by Gilbert. Although Robert Norman's work was well appreciated, it was not really in the learned scientific tradition, and Robert Norman was no kind of learned scientist. The following is a passage from his book where he defines his attitude towards science. He states why in his view an ordinary workman, writing in English, can do as well as a very learned, classically-trained scientist:

And albiet, it may be said by the learned in the Mathematicalles, as hath beene already written by some, that this is no question or Matter for a Mechanitian or Mariner to meddle with, no more than is the finding of the longitude, for that must be handled exquisitely by Geometricall demonstration, and Arithmeticall Calculation: in which Artes, they would have all Mechanitians and Seamen to be ignorant, or at leaste insufficientlie furnished to performe such a matter. . . . But I doe verily thinke, that notwithstanding the learned in those Sciences, being in their studies amongst their bookes, can imagine greate matters, and set downe their farre fetcht conceits, in faire showe, and with plawsible words wishing that all Mechanitians were such, as for want of utterance, should be forced to deliver unto them their knowledge and conceites, that they might flourish upon them, and applye them at their pleasures: yet there are in this land divers Mechanitians, that in their severall faculties and professions, have the use of those Artes at their fingers endes, and can applye them to their severall purposes, as effectually and more readily, than those that would most condemne them.*

William Gilbert

In fact, this is precisely what happened. There was a very learned man, William Gilbert, who was physician to Queen Elizabeth, and he studied this matter and wrote a book on it, which was really the start of the whole science of magnetism. It is also referred to in a much later book, Newton's *Principia*. Gilbert's is the first really scientific book dealing with a scientific problem and it proceeds in the proper form. Chapter I is on the writings of the Ancients on the subject of the magnet, the discovery of the lodestone. In the Preface he makes a plea for natural philosophy, which I think I should quote. Incidentally, this is translated: it was originally written, very appropriately, in Latin:

This natural philosophy is almost a new thing, unheard-of before; a very few writers have simply published some meagre accounts of certain magnetic forces. Therefore we do not at all quote the Ancients or the Greeks as our supporters, for neither can paltry Greek argumentation demonstrate the truth more subtilly nor Greek terms more effectively, nor can both elucidate it better. Our doctrine of the loadstone is contradictory most of the principles and axioms of the Greeks. Nor have we brought into this work any graces of rhetoric, any verbal ornateness, but have aimed simply at treating knotty questions about which little is known in such style and in such terms as are needed to make what is said clearly intelligible. Therefore we sometimes employ words new and unheard-of, not (as alchemists are wont to do) in order to veil things with a pedantic terminology and to make them dark and obscure, but in order that hidden things that have no name and that have never come into notice, may be plainly and fully published.**

These two manifestos published about the same time,

*Robert Norman, *The Newe Attractive*, London, 1581.

**W. Gilbert, *De Magnete*, 1st edition, 1600, translated by P. Fleury Mottelay 1893, Dover, New York, 1958.

slightly different in form, are really in the same spirit. What Gilbert actually did was to work out experimentally, the logic of a terrestrial magnet; in other words, to make a little earth, *terrella* he called it, and to put little compasses all round it to show that you had what we would now call a magnetic field, and this magnetic field determined the direction of the little compasses. He pointed out that the strength of the field is greater near the poles:

> . . . the supreme attractional power is at the pole, while the weaker and more sluggish power is in the parts nigh the equator. And as in the declination it is seen that this ordering and rotating force increases as we advance from the equator to the poles, so too does the coition of magnetic bodies grow stronger by the same degrees and in the same proportion. For at points remote from the pole, the lodestone does not pull magnetic bodies in a right line towards its centre, but they tend to it obliquely. . . .*

He also mentioned such elementary things as that the force is greater in a rod than in a sphere, showing that he understood the geometry. But the essential feature is that here is the experimental treatment of a subject. What is interesting, and which we come to much later in the history, is that after Gilbert it is not until the time of Coulomb, practically at the end of the eighteenth century, that there is any improvement in the knowledge of magnetism.

But Gilbert did not concentrate on magnetism, he went further. He noticed that there were other kinds of attractions, very much smaller, weaker attractions for which, instead of using a magnetic needle, he used a small mounted straw and found that it would turn around according to the effect of the attraction of other bodies. He deals with many imaginary effects, but he also deals with the real electric effect of amber when rubbed. The attractive power of amber was known, but it was Gilbert who first showed that it had the same kind of attraction as the lodestone, that it followed the same rules, although there was no direct connection because it did not particularly affect iron, but it affected everything regardless, although it was very weak. Nevertheless, out of that very weak thing grew the enormous strength of the electricity of modern times, but in a rather peculiar and roundabout way.

Ibid.

First of all, I should say something about how Gilbert used his own discovery in magnetism. He was not at all satisfied with the earth-bound properties of magnets; he wanted to find the operation of the magnet in the celestial spheres and he came to the conclusion that there was a general pervasive magnetism. The earth was a magnet and probably all the other planets were magnets; the sun would be a magnet and the whole process of the solar system, the Copernican system, was really a kind of magnetic engine, in which everything was held together by magnetic forces. In that sense he used it to buttress up a reasonable picture of the solar system; this was in 1600, before even Kepler or Galileo had got to work. He was completely wrong-headed as far as the particular forces were concerned. He believed that the magnetic forces actually drove the planets along. They are not driven by magnetic forces. Yet he was rather in advance of the times, because nowadays we recognise magnetic forces in the sky but they are not associated with solid bodies but with clouds of magnetic dust. The point is that Gilbert saw everything as magnetism and for that reason Bacon attacked him on the grounds that here was a man so keen on his subject that he made everything in the universe depend on that subject.

Electricity and the vacuum

I will now pass over into the field of the vacuum. In the seventeenth century, people were very interested in various kinds of luminescent phenomena. They confused them all, but they called all light-bearing phenomena the same thing: phosphorus. There was the common-or-garden phosphorus, which was found when fish went bad, or, occasionally, in toadstools, rotted wood and other things which shone in the dark. That was the ordinary element we now call phosphorus, made about that time by Kunkel from bones. It slowly burnt by itself, giving a pale, cold light. But the phosphorus which concerns us here is a different kind of phosphorus. Once they had a barometer, it was noticed that when mercury was shaken up and down in the barometer, it shone with a green light which disappeared when the shaking stopped. We say now, of course, that it is simply the excitation of the mercury vapour by frictional electricity. But in those days it was a wonderful

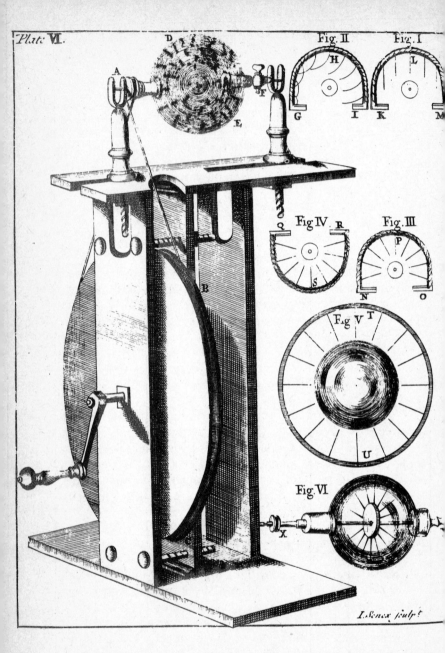

106 Rotation of an evacuated globe by an adapted spinning wheel; from Hauksbee, 1714

and interesting thing and it seemed to be – and indeed was – related to another kind of phosphorus, a very natural one, the Northern Lights, which seemed to have these rather vague and indefinite appearances, a meteoric phenomenon. A man called Hauksbee studied this; Hauksbee was Newton's laboratory assistant, although it is not very clear what he did for Newton. He certainly, however, made the first machine of the kind I will describe.

Fig. 106 shows a globe which is evacuated and it is rotated simply by an adapted spinning wheel. You spin it very fast, put a cloth or something similar on it with your hand and when you look inside, it seems to be lit up there. This was a very impressive sight and completely incomprehensible; I doubt whether anyone even now could explain it in any detail. It was one of those show experiments which people tried over and over again. They noted, also, that it did not really matter whether the globe had a vacuum in it or not; something happened even when it did not. When there was no vacuum it did not light up, but there were sparks and tinglings and people's hair stood on end – and various other queer phenomena of that sort occurred. It was explained in terms of what was called electric effluvium or fluid. A fluvium is simply another name for a very vague idea of some special liquid, which was supposed to come out of the globe and go off in various directions.

Stephen Gray

It was not until much later, about the time of Newton's death in 1727, that this problem was taken up and a very classical set of experiments were undertaken by Stephen Gray. Now, Stephen Gray, it has recently been discovered, was himself a pupil or student of Newton's, so the Newtonian experimental tradition went right on. It has been argued that the Newtonian experimental tradition in optics and electricity was more important to the development of physics than his purely mathematical and great discovery about gravitation, but I do not think this can be seriously maintained.

The problem, as Gray saw it, was just one of these straight through researches. His work is worth studying and you will find the best account of it in a book that is itself not only a classic of the history of science, but one that contributes very much to science in a roundabout way: *The History and Present State of Electricity*, by Joseph Priestley, 1775, his first work, apart from some theological ones. Priestley had been set on this task by Franklin who, as we will see (p. 288), himself had a great deal to do with electricity. There was no history of electricity and Franklin, who was much older than Priestley, suggested to him that he should write such a history. Priestley began the work, having an unusual idea about the writing of this scientific history – something which very few people would try nowadays, for the book is full of experiments. He said that in order to write a good history of science one must repeat all the experiments and sort out for oneself whether they were true accounts or not because it was not possible to tell from the author's accounts. Well, he did the experiments and, as you will see later, he got into considerable trouble about them.

But to go back to the actual story. Gray found that if you took a piece of glass – this was the extent of his apparatus, just a piece of glass tube open at each end – and rubbed it, the ends would attract light things. He found that if a cork was put in the open end, the cork attracted. He wondered what would happen if he went further and stuck a nail in the cork, and he found that the end of the nail attracted. He put a little metal nob on the end of a metal rod in the cork and found that that attracted. The metal rod was made longer and longer and it still attracted. In time the rod became too long and began to sag. He then tried hanging the rod from the ceiling with pieces of string but this did not work. Then – why he thought of this is not clear – he decided to use a piece of silk instead of string to support the rod, and this worked. Then, as he had no more metal, he tried some ordinary hemp thread, stretched out and hung on silk loops. He was spending the summer – it must have been a very good summer – with a friend in the country who had one of these nice Elizabethan houses with a big gallery all along the house. They ran this thread on loops from one end of the gallery to the other and it worked right through the gallery. He decided to turn it round and bring it back again – it still worked. Then, immensely daring, they took it out in the garden and they found that it worked all round the garden.

Gray then realised that he had made, not only a discovery of the way electricity could move, but really an electric telegraph which, he suggested, could be used for sending messages if the proper way of supporting it could be found. Then the weather turned damp and nothing worked any more. But he had thus established the first principle which was that electricity was something that would travel. He could distinguish between two things which he called electrics, that is, things that generate electricity, and non-electrics, that is, things that electricity would go through – what we would call conductors. He found that metals were the best conductors, that water was a conductor, and so forth. All that in just about a month's work – all that we do in elementary classes in physics! But at that time, it was new: no one had ever done it before.

The next person to take up the subject – they were all writing letters to each other and there were various publications – was a kind of official scientist in France, C. F. Dufay, who was responsible for dyes and the chemical industry which, in about 1740 or so, was beginning to flourish. He made the startling discovery that there were two kinds of electricity, one of which is obtained from rubbing a piece of glass, and the other from rubbing a piece of resin. He found that things could be divided into two categories, vitreous electric and resinous electric, and the two were antagonistic to each other. The same kinds of electricity repelled each other, but opposite kinds attracted.

The Leyden Jar

The fun went on, but it was still a game for scientific amateurs because, first, it was no use and, secondly, it did not make sense, it did not fit in with any of the mathematical theories of the time. Nevertheless, pieces of apparatus were made and passed around and people entertained each other with them in country houses. People in those days had much less to do and they even took up scientific experiments for entertainment to while away the time and to enjoy as parlour tricks at parties. That was all it was to start with. But it soon had a big lift because one or two people had bright ideas about it; first von Kleist and then Musschenbroek at Leyden. Musschenbroek was a new kind of person, a scientific-instrument maker. There had been optical-instrument makers before – you will remember I pointed out that Watt was an optical-instrument maker – but

Musschenbroek made general pieces of scientific apparatus. He had the bright idea that it might be possible to store the electric fluid and that the obvious place to store it was in a bottle. He took a bottle and half filled it with water and put the electric conductor into it – but nothing happened!

Fig. 107 illustrates this experiment. It is a typical eighteenth-century picture. There is the scientist, dressed like a gentleman, with his operator dressed like a valet – which he probably was – and there is the bottle and the machine. The machine is essentially Hauksbee's machine and there is a boy somewhere off in the background turning the machine. *A* is something they called a prime collector, which is a metal bar; there are actually two in this picture but two are not necessary. The operator held the bottle in one hand and nothing at all happened. Then he touched the prime conductor and he received the first known electric shock. Many people had received electric shocks before, either in lightning storms – usually fatal – or from a torpedo, the electric eel, which is a pretty tough shock, too, having low voltage with very high amperage. But this was the first time a person really knowingly received a shock: it frightened Musschenbroek to such an extent that he said that not for the kingdom of France would he undertake such an experiment again. Nevertheless, out of mere curiosity, of course, he undertook it again the very next day.

Once he pursued this idea, electricity 'came to town' in a very big way because now it was possible to produce real effects. First, these shocks could be produced, and these were quite new; then, sparks could be produced, not just miserable little sparks that could just be seen in a dark room, but big sparks that crackled or could be seen in broad daylight, and everyone took up electricity and it became very fashionable. What is more, they discovered something very elementary, that is, if two people held hands they could both receive shocks simultaneously. Among ecclesiastics trying it out, one abbot gave a shock to all his monks at one time and the King gave a shock to the whole body of his Grenadier Guards and made them all jump in absolute time! It was not only just very exciting, doctors began to use it. There was a tendency in medicine to believe that any new phenomenon cured all diseases. Gradually they would discover that perhaps not this

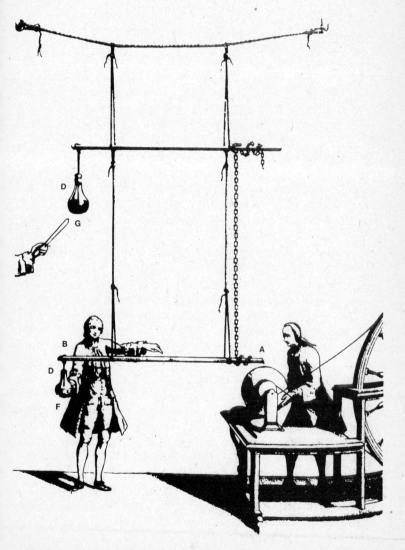

107 The gentleman scientist: Musschenbroek's experiment

or that disease could be cured but there was always a residue that responded and there are still such residues; people being treated one way or another by shock, I think rather misguidedly, for mental defects. But, in any case, electricity had really arrived, everyone knew about it and talked about it.

Benjamin Franklin

It was at that point that there came the first contribution to science from the New World, through Benjamin Franklin. I have not the time to deal with Franklin's quite extraordinary career in any detail. He was very much a self-made man, a printer and editor at the age of twelve: he ran away from his home in Boston because the puritanical directorate there did not like what he printed in his paper. He went to Pennsylvania and there became a postmaster and printer and various other things, and, in his spare time, he interested himself in what was going on in science in Europe. He had visited England and became involved with scientists there in a most ingenious way. He took with him one little thing, a piece of asbestos, which would withstand fire. This he presented to Lord Macclesfield, the then President of the Royal Society, and he was admitted to all its meetings and allowed to meet the scientists there. The asbestos, incidentally, had another curious story attached to it. It was taken up by John Wesley, who postulated the possibility of eternal damnation because, as the asbestos could endure the flames for ever without being consumed, so the damned souls in hell could do likewise.

But to return to Franklin; he ordered a box of electrical equipment from Europe, which was sent to him in America and in even less time than Gray, he mastered the whole matter. First of all he discovered that it was quite unnecessary to have vitreous and resinous electricity, the presence of one ensured the absence of the other. He got the idea of positive and negative electricity, rather inconveniently for us, the wrong way round, but it had to be called one thing or the other. He showed that matter naturally contained an equilibrium, if some charge were taken away, you charged what you put it to positively and what was left behind negatively. He demonstrated how the Leyden jar worked and how the charges really resided in the glass and not in the electrodes. He then carried out his crucial experiment – crucial to the history of the subject – and he was very lucky to survive it. He charged a Leyden jar from a lightning discharge. Luckily he did not get much of the lightning into his Leyden jar. The next man who tried it, in St Petersburg, was killed. But not only did Franklin do it, he immediately turned it to practical use in the lightning conductor.

The lightning conductor

Lightning conductors are used in Britain but they are far more important in the United States, as anyone knows who has been there. You cannot touch a door handle there in the winter time without getting a moderately bad shock because the air is so dry and there are many electrical charges around. So Franklin's conductor really sold and, at the same time, it impressed people that electricity was useful, quite apart from the rather hypothetical medical uses it might have.

Unfortunately, just at that time, there was a certain amount of difference of opinion between the colonies and the government. Franklin was the agent for Pennsylvania in London and became involved in a great deal of trouble on this account. Poor King George III, who was the only really scientific king we ever had – Charles II was not really serious – had complete laboratories and observatories of his own and worked there. His views, however, became coloured more by political considerations than scientific ones when he would not use the lightning conductor because it had been used by 'that rebel Franklin', and he had it modified. As a matter of fact, the President of the Royal Society had to resign on account of it. I will just quote shortly from a contemporary verse on the subject, composed because the king had insisted that lightning conductors should have spherical tops instead of the sharp pointed ones proposed by Franklin:

> While you, great George, for safety hunt,
> And sharp conductors change for blunt,
> The nation's out of joint.
> Franklin a wiser course pursues,
> And all your thunder fearless views,
> By keeping to the point.

Coulomb and the law of attraction

Here someone else comes into the picture. You will have noticed that so far the whole of this theory of electricity is qualitative. Now quantity enters into it, in the first place with Coulomb who occupied an official position in what might be called the National Physical Laboratory for France, responsible for optical instrumentation, compasses and so forth. He was charged, as people often were from time to time, say, every two or three hundred years, with improving the compass, the compass not having been seriously improved since Gilbert's time. We are now talking of the 1780s. Everybody was in an improving state of mind and Coulomb discovered that no one knew precisely how a compass worked and what were the forces. No one had ever thought of measuring them. The curious thing is that Coulomb's real work before then had been in an entirely different, a purely mechanical, field. As he was responsible for the inspection of naval stores, he had to measure the strength of cables, of iron bars, and such like, and one of the aspects he had to measure was their strength in twisting, their torsion. So he invented the torsion balance to do what it was really supposed to do, measure how much an iron bar could be twisted before it broke. He soon realised that he now had something that could be used for measuring any kind of force and it was a very convenient way of doing it. So he developed the torsion balance with which he showed – and this is the first important thing – that the magnetic poles and the electric charges appeared to be concentrated at points in the Newtonian sense, and that the forces were proportional to the inverse square of the distance.

In other words, the three natural forces of gravitation, electricity and magnetism, all obeyed – quite properly as he discovered it – Coulomb's law. Fig. 108 shows the apparatus he built. You see it is the basis of a really scientific apparatus. There is the suspension, although this is not new (I pointed out earlier that the Chinese were using suspensions like that for compasses in the seventh century) but the idea of the measured scale, the scale is at the top, is entirely new. In this way Coulomb began to make a quantitative picture of electricity.

The next person to enter the scene was one of the most

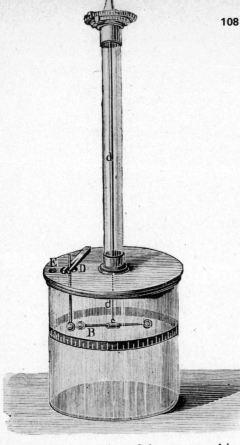

remarkable scientists of the age – and it was a very remarkable age – Cavendish. Cavendish, who had the misfortune to be a younger brother of an Earl, had an enormous amount of money. He was extremely unsociable and saw nobody. In the season, he would come into London once a week to meetings of the Royal Society, travelling in a closed coach so that he could not be seen. He was shepherded into the Royal Society and took his seat. He never communicated with people – except for about half a dozen or so – and apart from them he did not speak to anyone. He was pretty bad at publishing, too. Luckily we know that he did write because, years afterwards, Clerk Maxwell published all his papers, but at the time what he did was unknown. He made careful measurements and discovered originally all the laws of electrostatics, of shielding, even of the dielectric properties of materials, all the things which Faraday had to discover in the next century without knowing that Cavendish had already done so.

I cannot, for want of space, make mention now of all the things he did, but among them was his interest in animal electricity. Animal electricity was bound to have received attention because of the obviously physiological effects of the electric shock. But it was soon noticed that the effect worked backwards and John Hunter, the great doctor and anatomist, was studying these phenomena for which Cavendish made a model of an electric fish. Here, it seemed, was something that with one or other of its organs could generate electricity. So he made a model of a fish with Leyden jars placed inside so that they could initiate the shock of a fish even in water. The whole mystery of the electric fish makes a story of its own, and it is still not fully understood. It was not known until after the last war that fish could communicate with each other by means of telegraphic electrical signals and that they have receptor organs as well as emitters.

Galvani: animal electricity

The only kind of electricity hitherto discussed is static electricity. Now we come to the kind of electricity we know as current electricity. How it was discovered is a separate story. It was the result of the study of animal electricity that the next and very important step was made. It was also an accidental discovery. Fig. 109 is a composite picture. We are now at the end of the eighteenth century, in the 1790s, and there is a great revival of science in a place where it had been dead for some time, in North Italy. In the picture, you see the Professor of Anatomy in Bologna, Professor Galvani, doing experiments on a very nice convenient table. He is giving a combined lecture, it seems, on physiology and electricity; I do not understand how they were mixed up in this time-table but there are the physiological preparations. It was known that if you pinched the nerve of a frog's leg, it twitched. Incidentally, there was nothing very unusual about producing frog's legs for experiment, they were purchased for dinner round the corner, for the legs alone were eaten. To continue about the experiment, every time they produced a spark on the electrical machine at one end of the

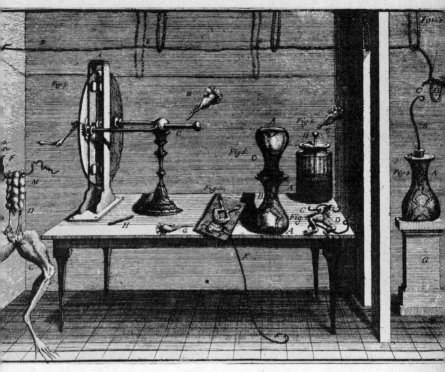

109 Galvani's experiments on electricity and physiology; from *The Effects of Electricity on Muscular Motion* (1791)

table, some bright person – or perhaps the Professor himself – noticed that all the frogs' legs at the other end of the table gave a jump. As a matter of fact, Galvani had discovered more than he actually knew, because this was not really a conduction effect, it was actually an electrical oscillation that had been transmitted. He had discovered radio without knowing it!

Galvani continued with many experiments of this kind until he discovered, again by accident, when he was looking for atmospheric electricity, that when a cloud passed overhead, the frogs' legs also jumped, when the static electric air potential gradient altered. He noticed, further, that as he hung the legs on an iron railing, when the frogs' legs attached to a silver hook touched the iron balustrade, they jerked. He then spotted that all that was required was two different metals and the frogs' legs were used simply as a 'galvanometer' – he simply measured something that was happening with the frogs' legs. The interesting thing is that for a long time it was considered that these effects were in the realm of physiology, that new things were being dealt with, not ordinary electricity. Neither the electric eel nor this seemed to fit in with the laws of electricity because, when something was wet, you could not maintain any potential, there were no sparks and so forth; it was a quite new phenomenon so they called it animal electricity. It was said that this proved there was organic and inorganic electricity.

Volta: the electric battery

This infuriated Galvani's colleague, Volta, Professor of Physics at Pavia. He absolutely rejected the idea of animal electricity because he had already done some very straight experimenting in static electricity. Volta had discovered the electrophorus with which, with a relatively low charge, he could load up a Leyden jar to as much as he wanted. He also knew that by using the electrophorus, which was really a multiplying machine, he could detect what we would now call very low 'voltages' of electricity. He would multiply the very low voltages until they were large enough to work an electrometer. It was a very complicated way of measuring things but the only way, because an electrometer would not work for a tenth of a hundredth of a volt. So first of all Volta established this method of multiplying voltage and from it he perceived that the voltage in Galvani's experiments was due to the different metals used. He had already shown that there was a contact potential between different metals when they were put together and he wondered whether this animal electricity was just the effect of this contact potential.

He went a little way to meet Galvani by using something wet, that is, by putting some wet blotting paper between his discs and in that way he soon produced the Voltaic pile. A pile was necessary because Volta's only way of measuring electricity was by getting enough potential to raise a spark. He had to put twenty or thirty of these discs on top of one another to get his first battery. Once he had done this he realised that here was something which had nothing to do with the animal in the first place, the animal was not necessary.

Galvani scored off him after that – it just illustrates the power of divergence in science – by showing that the metals were not necessary either! Galvani was able to use salt solutions of different strengths and made a circuit in which the elements were entirely physiological. When the nerve was excited in any other way one could be made to stimulate another twitch. There are, of course, potentials generated in contracting muscles and, in fact, the big potentials which are generated by the electric fish are generated by a very ingenious perversion of muscle action potential in what is called an end plate. Each muscle fibre cell has an end plate to which a nerve goes. It is a very complicated process: the nerve has to generate a chemical which then excites the end plate material which, in turn, excites the muscle. In the electric organ of the eel there is an end plate which develops very much and only has very little muscle attached to it so that electricity is generated instead of motion and that is what electric organs are – degenerate muscle. But that is a different story, that is physiological electricity. The main line of investigation went right through Volta to the next school of current electricity which began, again, mainly in England.

The Royal Institution

We have now crossed the century and have arrived in the nineteenth century, 1800 or so. Thompson, Count Rumford, whom I mentioned earlier in connection with his work on heat (p. 275), was able to set up in 1799 the first all-round scientific institute in the world, which is still there, working in the same place where he first set it up – the Royal Institution. It was intended for mechanics and others who were to be taught about inventions. It was, however, not very successful in this because the gentry did not like mixing with the mechanics. At first they arranged a way by which the mechanics could go up by a back staircase to the gallery to avoid their having to meet the gentry. The latter took the opportunity of Rumford's absence, however, to close this up and the Royal Institution became a really exclusive place. They chose a very bright and fashionable young man to run it – Humphry Davy. He was probably the finest lecturer of his time; he always produced the most wonderful effects in his lectures, especially, as we know now, on the chemical side. But one of the most impressive demonstrations he gave was with a really big battery. He tried it out and first produced all the effects that had been obtained with the battery of Leyden jars. The word 'battery', incidentally, was first used, and most properly used, for Leyden jars; it was used only later for current electricity batteries. Davy then showed that you could use batteries to decompose metal salts. He developed electrolysis and this was to lead – though I have not the time to discuss them – to the great electromagnetic discoveries of the nineteenth century.

Just to introduce the subject, I will now show what are the kind of functions that make for scientific discoveries. The quantitative functions are a large field of interest, that is, it is necessary to involve a great many people. If you were to try to read – I have at times, in fact, made it a principle to do so – all the second-rate books on science produced through the ages, not the first-rate ones because they were written by geniuses and were never understood at the time, but books written by ordinary people and understood by ordinary people, it is clear that most of what is in them is nonsense. Many people were interested in electricity and all thought that something was bound to come of it. They were looking for a natural philosophy which would join everything together. It was appreciated that there were many forces in nature, one of them was gravitation and there were electricity and magnetism. They must, it seemed to them, be related to each other. At first the various forces obstinately refused to relate themselves to each other. Then came the celebrated experiment of Oersted in 1820, when a battery and a coil, on the one hand, and a compass needle, on the other, had been left on the lecture table together. When the current was turned on it was found that the compass needle turned and so the effect of the electric current on magnetism was demonstrated. This led directly to electromagnetism and indirectly to the electric telegraph.

It was another ten years before the converse effect was discovered and by that time everyone expected that there must be a converse effect. The converse effect was discovered by Faraday, also at the Royal Institution. Michael Faraday, the greatest physicist of the nineteenth century and possibly the greatest experimental physicist of all time, had been picked up by Davy as an assistant and soon showed himself to be a better scientist than his master. It is to him we owe the proof of the relations of the various physical forces to each other.

For a long time people just put their magnets near the wires carrying currents and left them there but found that nothing happened. The idea that it was necessary to move the magnets or the currents, the dynamic aspects, was entirely Faraday's. And it was not an accident, it was very carefully worked out as can be seen from Faraday's notebooks. He showed how mechanical work can be transformed into electricity, and therefore brought electricity into the scope of the rising mechanical-industrial revolution.

The lag in electrical development

This connection should really have led immediately to a major electrical industry but, as I have shown in one of my own books,* it took a very long time to develop. The reason was largely economic. It would have been possible to have produced electricity in a big way in 1830. Faraday had already made an electro-magnet which gave quite considerable current, but no one wanted it. Electricity could not be sold because there were no buyers and you could not use electricity because no one could make it until much later. Things developed in a tortuous way. The first call for electricity was through fashion: people were becoming moderately rich, not rich enough to afford silver spoons, but what about having electro-plated spoons? For that a good source of current was required and the magneto machine of Faraday, slightly improved, worked very well for this purpose. Then it was used for where really bright lights were needed, arc lights, and for lighthouses. Gradually the uses increased and as the demand increased production increased.

I have omitted to talk of one aspect, the telegraph, which really started it all, but I cannot go into that in detail now. There were really two aspects of the utilisation of electricity, one which might be called the light-industry aspect, the other the heavy-industry aspect. The light-industry aspect is simply communication, and the means of communication was to be the telegraph. Fig. 110 is of a telegraph and a very old-fashioned one at that. It is an ordinary signal telegraph and the houses are still there. There are a set of hills between London and

*Science and Industry in the Nineteenth Century, London, 1953.

CODE FOR SIGNALLING

A	B	C	D	E	F	G	H	I	J	K	L	M
N	O	P	Q	R	S	T	U	V	W	X	Y	Z

110 The semaphore Telegraph of Chappe

Portsmouth and these telegraph houses can still be seen on them. But the heavy electrical industry first concentrated on the production of light. For that it is necessary to have a means of making cheap lamps of a few candle-power rather than arc lamps for lighthouses. So it depended on the development of the filament lamp and the power to run it.

Fig. 111 is a picture of an early power station in London. They were already showing an electric, platinum filament lamp at the Royal Institution in 1800, yet the first filament lamps in practical use only came in in 1830 and in that century it was quite a feat to bridge such a small gap as between the arc and filament light but, again, the gap is essentially functional. At first arc lights were used but no one wanted an arc light in his dining room, that would really have been a bit too much. They were fine in some main streets or in harbours or lighthouses. So the arc light never really caught on. The problem was what is called the division of electric light, a problem with two sides to it: first of all how to make a small

111 Metropolitan Electric Supply Company's high voltage power station, Sardinia Street, 1896: one of the first power stations in London

electric light and then how to make it worth while to get current on this small scale. Until these two problems were solved – the first by Swan and the other by Edison – electricity was not a practical proposition. Swan had the right idea for better lighting but, being English and brought up with the proper respect for the best people, he wanted to make electric systems for noblemen's houses. Obviously, every nobleman who could afford to have electric light could also afford to have his own generating station in his own backyard, or worked by a waterfall in the park. But Swan did not make much progress with this. Edison, on the other hand, who had already put in some very good work fixing up ticker-tape systems all over New York, disregarding every franchise, realised that it was possible to combine a distribution system with a power station that provided electric light to many houses. And that was the beginning of the modern age.

12

Conclusions

The first eleven chapters of this book are taken from a recording of lectures given during the Michaelmas term of 1962. They constitute substantially the course of these lectures given between 1946 and 1962. This last chapter is not in fact a lecture delivered to the course and it is added to complete this book and bring it up to date.

The lectures were intended to deal with the history of physics in its earliest phases, that is, classical physics before the many great transformations which began in 1896 with the discovery of X-rays and the modern atomic theory. As can be seen, this plan was not quite carried out. The time was mostly spent on discussion of the Renaissance and the seventeenth century which ushered in modern science, from the work of Copernicus to that of Newton. In the later chapters, 9, 10 and 11, an attempt was made to pack in the rest of physics – an understanding of heat and its mechanical equivalent, exemplified in the development of heat engines. Then followed a chapter on optics and, finally, the whole subject of electricity and magnetism was sketched into Chapter 12. The great work of the nineteenth-century physicists, the time in which physics really took shape, was passed over very quickly, and the greatest physicists of the period, Faraday and Maxwell, were hardly mentioned.

This chapter, then, attempts to carry the story further through the nineteenth century, not as a consecutive history but rather as a piece of hindsight, irrelevant to the ideas of the nineteenth century but concentrating on those parts of the subject which carried the physicists forward to the breakthrough at the end of the century.

I began Chapter 11 with a bald statement of the greatest developments of the nineteenth century itself, which can be summarised in two major discoveries: the interconvertibility of all forms of energy, the first and second laws of thermodynamics; and the electro-magnetic theory of light. By the end of the century, these between them seemed to give a self-consistent, coherent picture of the universe, both as to its origin and its fate. Although the basis of both of these discoveries was firmly laid in experimental science and in technology, they are essentially theoretical constructs, a matter of doing the right experiments and thinking about their applications, and they must be considered as a combination of experiment and theory which has been the essence of physics almost up to our own time. There is undoubtedly a view now that the experiment is really a kind of unfair physics and that if we only thought hard enough we would have all the answers without bothering to experiment – a view expounded by the great theoretical physicists Planck and Einstein, through to the elementary particle physicists of our time. Nevertheless, it is still generally considered not only important but vital to continue with experimentation, even though it has now become so expensive as to threaten the whole future of physics.

There was also, in the nineteenth century, a radical change in the very status of physics or, more accurately, in the nature of the statements about physics. In the nineteenth century, physics was concerned with statements of fact rather than with explanations. It really consisted of a catalogue of the properties and behaviour of matter, dealing with *how* matter behaved not *why* it did so. It contained a number of constants which perhaps could be reduced to some extent to a comparatively few fundamental concepts, such as Avogadro's number or the quantum of action; but much brute fact was left over to answer the questions as to how things behaved the way they did but not as to why they should do so. Water froze at 0°C and boiled at 100°C, and it was improper to ask why it should melt and boil at these temperatures. I have chosen this example because the answers are so conventional that they are really truisms. 0 and 100 have been chosen as standard temperatures because the properties of water are very convenient as a reference substance, although actually water is a highly complex liquid with unique properties, and it would have been profitless to ask for an explanation of its properties until well into the twentieth century.

As these lectures were supposed to be concerned essentially with experimental physics and not with theoretical physics, I had arbitrarily to limit what they contain. But I will try to represent in this chapter the major lines of action and thought that went into the creation of modern physics, which attempts to provide something of the *why* of physics as well as the *how*.

The inter-relationship of natural forces

The legacy of eighteenth-century physics was the existence of a number of quantitatively determinable forces, each with its own peculiar quality, of which we still have the relic in the names of the different sub-branches of the subject – heat, light and sound, electricity and magnetism. The first task of the nineteenth-century physicists was to show the relation between these forces. To begin with they were all studied together, as I have already indicated, and hence their relations were practical. Those of electricity and magnetism, achieved by the experiments of Oersted and Faraday, were the first to be unified, with tremendous consequences in industry, giving birth to the electrical industry, the first purely scientific industry.

Now, Faraday was a magnificent experimenter but not strong in mathematics. His work was complemented by that of Maxwell, no negligible experimenter himself. It was out of the relationships found by Faraday between the various forces that the concept of electro-magnetic fields arose. They had been anticipated as far back as the days of Gilbert, whose little compass needles around his model earth or *terrella* showed how the magnetic field could now be made visible to every school-boy who scattered iron filings on a piece of paper near a magnet. Faraday called these rows of filings magnetic lines of force; they were analogous to the electrical lines of force formed by electric charges. By an extremely ingenious exercise of imagination, he was able to discover the principles of lines of force and to put the whole of his observations in terms of lines of force. His discovery of the generation of electric currents by magnetic fields in motion could be expressed in terms of lines of force cutting each other. With a permanent magnet the lines of force necessarily formed closed loops, with electric lines of force they necessarily ended on electric charges.

Other workers, particularly in Europe, like Ampère and Gauss, measured electric quantities, but the theory in its full sense was made mathematical only by Maxwell who translated Faraday's pictures into quantitative terms in differential equations. Four principal equations were sufficient to summarise the mutual behaviour of electro-magnetic fields. Maxwell's equations were able to do far more. Electrical magnetic quantities can be expressed in absolute terms of unit poles and unit charges. They were in fact two systems of units and the ratio between them turned out to be a velocity of 3×10^{10} cm./sec., an astonishing velocity which appeared to be at the same time the velocity of light. This was to be the basis of Maxwell's great generalisation, the electro-magnetic theory of light, published in 1873, which united at once the field of optics with that of electricity and magnetism, and so led later, through the work of Herz, to a concept of light waves being originated by oscillating electric charges, the basis of radio and the whole of modern wireless communication.

Optics itself showed enormous developments in the early nineteenth century. Young in England and Fresnel in France, showed the immense advantage of Huygens's theory of light and of vibration – the wave theory of light – and Newton's particle work seems to have been superseded. About the same time Frauenhofer showed experimentally that the solar spectrum was not, as Newton had thought, continuous, but was crossed by black lines corresponding to radiations which were absorbed in the surface layers of the sun. Almost immediately it was found that these same radiations were produced by excited atoms of gas, as when salt is dropped into a flame, and we are now familiar with them in our sodium and mercury street lamps as well as in all fluorescent lighting. Fluorescent cold white lamps have largely superseded the old hot orange filament lamps.

At first this knowledge was essentially of chemical and astronomical use, identifying every element by its spectrum and showing that all the bodies in the universe consisted of the same elements. Spectral analysis was now becoming an important and delicate chemical tool. New elements like helium were discovered by its means. Partly as a result of this, the elements themselves, of which more and more were being discovered during the nineteenth century, were put in order by the Russian chemist Mendeleev. A periodic system of the elements was described but could not be explained until the twentieth century.

Meanwhile, Faraday continued his work in two directions which were to be vitally important in the future. He showed further connections between electricity, magnetism and light; the polarisation of light was shown to be affected by magnetic fields which rotated the plane of polarisation as did some natural crystals like quartz and organic chemicals like sugar. In the mid-century, the distinguished French chemist, Pasteur, showed that such chemicals found in nature were the fruits of asymmetric forces formed by living organisms, and this was the beginning of his work on bacteriology which was to have such an enormous effect in the fields of medicine and agriculture.

Returning to physics: Faraday also investigated the phenomena that had started off the study of electricity – the luminous effects of the electric discharge in a vacuum (p. 283), or, rather, in gases at very low pressures. This was the beginning of a series of studies carried out by Crookes and others, that were to culminate in the discovery of X-rays by Röntgen, and of the properties of the carrier of electricity, the electron, by Thomson. Faraday had previously shown that electricity could be carried through liquids in unit quantities, the ionic charge 'e', the same for all elements but, on account of the prejudice against atomism, no one chose to attribute this to any new particle.

The conservation of energy

Now we come to the second great unifying idea of the nineteenth century – the conservation of energy. This was to appear around 1850, rather late, considering the advanced state of science at the time, and it did not derive from among the now professional physicists but essentially from the work of outsiders, engineers or physiologists. The practice of heat embodied in the steam engine and later in the locomotive had run ahead of the theory. The amount of work that could be got out of the same amount of fuel seemed to be always increasing. Improvements in the type of engines, which, to start with, were of an extremely low efficiency, could be increased ten-fold or more by *ad hoc* methods. Was there any theoretical limit to this?

Sadi Carnot considered the problem of efficiency as one of lifting the energy from the hot boiler and discharging it into a cold condenser, a problem like that of the flow of water. He argued that there was a limit to this, that the amount was limited by the difference in temperature between the boiler and the condenser. Thus the Second Law of Thermodynamics, that of the availability of heat, was discovered before the first one on its mechanical equivalent, although later Carnot perceived that what was in effect transferred in the heat engine was not so much caloric or quantity of heat as entropy, and was driven to recognise that a given amount of work always corresponded to the limit of the amount of heat available. He had established the First Law of Thermodynamics in 1830, but his notebooks were not published until 1927 and he got no credit for it. Meanwhile, the idea of the conservation and transformation of energy was so much in the air that it was impossible to say who finally discovered or published it. Credit is usually given to a German – Dr J. R. Mayer – who observed, as a ship's doctor in the East Indies, that not so much heat was required to preserve life there as in the north. He made some rather crude experiments to show this, finally seizing on the principle that heat and work were both examples of a common quantity of energy which could appear in many different forms and could be exchanged at a fixed rate.

This was the mechanical equivalent of heat. It was afterwards measured accurately by Joule in Manchester. This work was for long ignored by the professional physicists, because they did not find it in Newton's work. Joule had come to this study in order to determine the efficiency of electrical motors in relation to the amount of metal used up in the batteries. Afterwards, the idea of the conservation of energy was taken up by the regular physicists like Helmholtz and Thomson, later Lord Kelvin, and the idea of the universality of energy was made a corner-stone of the general view of physics.

According to the Second Law of Thermodynamics, however, the energy of the universe – which contained vast hot and cold regions – was bound to even out in a universal tepidity. The conception of a running-down universe was rather depressing, because at the same time it indicated that the universe had only been going for a relatively short time and could only last for a correspondingly short future, until universal heat-death overcame it. This pessimistic outlook was in complete contrast to all that had been put forward by the biologist Darwin, whose theories required a much longer time for the evolution of animals and plants. In a negative way this presaged, although it did not actually lead to, the physical revolution, demanding new sources of energy which were to be found in nuclear transformations from which the sun derives its heat.

The idea of the universality of the conservation of energy had at the time – in the middle of the nineteenth century – a very mixed influence on the programme of physics. Energy seemed to be a way of avoiding the conception of the atomic nature of chemical reactions. At this point it is necessary to state that the progress of chemistry at the end of the eighteenth century and the beginning of the nineteenth had been so great that it dominated the whole of physical science. Priestley's and Lavoisier's pneumatic revolution helped in the establishment of a rational chemistry and the basic idea of oxidation as the source of energy not only for combustion but for the metabolism of living organisms. Later, Dalton, himself a physicist and meteorologist, deduced from some of these gas reactions that they could be explained best by the hypothesis of atoms, essentially the same as those of the Greeks, but themselves of definite and different characters, each having its own weight and combining in integral proportions with the others. When atoms combined, a certain amount of heat was given off, the heat of combustion, which was the source both of industrial and biological motion. In spite of the logical success of the atomic theory, the very small size and invisibility of the atoms, together with the prejudice inherent in the prevailing religious ideas that they were in some way materialistic, delayed its acceptance until right up to the end of the century, particularly in the fields where the chemist's balance could not be used, as in that of electricity, or which were not understood, as in Faraday's work on electrolysis.

The concept of the atom, however, as an unalterable piece of matter remained, and it was not until the discovery of radioactivity by Becquerel, the Curies and Rutherford, at the turn of the century, that the modern concept of the atom composed of fundamental particles could be established. In a sense, this conflict between the continuum and the atomic nature of things was a continuation of the struggle in the seventeenth century between Descartes and Newton. It is still going on but has passed to the higher field of mathematical physics.

The roots of modern physics

In general, the development of physics in the nineteenth century was a magnificent and coherent effort, encompassing whole phenomena in a few experimental generalisations. Yet these were not to lead, as their proponents hoped, to a straight, all-embracing picture. The progress of science was to be otherwise. It was where the phenomena obstinately refused to fit in with the theories that new work and new ideas had to enter in a revolutionary way. These are the roots of modern physics. I shall not attempt to explain them completely now because, in the plan of my lectures, this was to be done by other teachers in their actual discussion of modern physics, both theoretical and experimental; the two approaches are so bound up with each other that they cannot be separated.

We may begin with optics and electricity. The theory of electricity in terms of electrons raised questions which were not satisfied by Maxwell's equations. This led to the work of Einstein, beginning with the new principle that it was impossible to determine the absolute velocity of an isolated object; only its relative velocity. This concept of relativity was to create a totally new attitude to physics and to establish the velocity of light as a universal velocity that could not be exceeded anywhere in the universe. The first developments led him further to the general theory of relativity and the real equivalence of gravity with accelerating motion, subjects which cannot be dealt with in this book as they belong essentially to the science of modern physics.

There was another jumping-off place in nineteenth-century physics in the theory of heat and its distribution of energy in the spectrum. According to the older ideas, a hot body lost heat in such a way that the greater part of the energy was to be found in the wavelengths of high frequencies. This was a consequence of the idea of heat being lost in a continuous way. It was Planck who put forward the daring and even absurd hypothesis that heat could only be lost in actual separate packets, the quanta of action, and, thus, most of the energy would be concentrated in the low frequencies, as was found to be the case. This idea was equivalent to showing that not only electricity but energy itself was atomic and not continuous. It was a further instance of the triumph of the atomic over the continuous view of the universe. It was to lead to the quantum theory, which provided the explanation of many of the obscure effects of chemistry, the concept, for instance, of discrete energy levels of atomic processes. The completion of this idea was to come from a study of the spectral lines themselves. The fact that there were definite frequencies and not continua showed that the atom itself must have a very complicated and particulate structure. This had already been guessed by Balmer in 1880, whose analysis of the simplest spectrum – that of hydrogen – was one electron per atom, showing it to consist of definite terms in which each special line was to mark the difference of the energy between one term and another. This could not be sufficiently understood until Bohr's theory of the atom appeared, that is, not until 1913, and with it came the key to the vast complexity of all other atomic spectra on earth or on the sun and stars.

The new era in physics

It was, however, from the study of X-rays that the real clue to the new physics was to come. The fact that electrical discharges gave rise to fluorescent phenomena, discovered by Röntgen, led almost accidentally to that of similar fluorescent appearances produced spontaneously from certain minerals, in particular, those containing uranium. This observation, first made by Becquerel, was to lead to the discovery of radioactivity and the impermanence of atoms. It is somewhat ironical that the atom was first accepted just at the time when it was demonstrated as impermanent. The explanation was to come in the first place from the experiments of Rutherford and was then established by Bohr. This gave rise to the whole of nuclear physics of today, including atomic disintegration and the principle of indeterminacy: the picture of the atom as a miniature solar system containing a highly charged and heavy nucleus surrounded by a planetary system of electrons. This concept, which had first appeared in an entirely hypothetical and analogical form in Newton's work, a new paradigm, was to prove the beginning of a new era in physics, and its application in the form of nuclear energy and the atom bomb was to transform the whole of the prospects – good and evil – for humanity.

These are the main characteristics of the breakthrough, showing how they led up from the classical physics of the nineteenth century to the present.

Index

physics, definition of, 15; social aspects of, 22–3; modern, 32; atomic, 32–3; Greek origins of, 83–103; heliocentric, 132 ff.

Picard, J., 223

Pisa, Leaning Tower of, Galileo's experiment on the, 180; Cathedral, 183

Pisistratus, 88–9

Planck, Max, 208, 302, 308

planets, and the elements, 81; movements of, 96–7, 156–9, 160 ff., 238; see also solar system

Plato, 31, 81, 89–91, 166

plough, 123

Plutarch, 101

pneumatic catapult, 64

pneumatics, 61–4; see also pumps, vacuum

Pole Star, 98

Polynesians, 46

Pompeii, 55

Portugal, navigation in, 153

power stations, 298–9

presses, 55, 57

pressure cooker, the first, 255

Priestley, Joseph, 43 n., 307; The History and Present State of Electricity, 284

prism, the, 218–21

propulsion, blowing as a form of, 63

Prussian Tables, the, 160

Ptolemy, 92, 112, 150, 152, 156

pulsilogium, Galileo's, 183–4

pumps, 147–9, 240 ff.; vacuum, 241–55; pulsometer, 258–60

pyrites, see firestone

Pythagoras, 75–7, 80, 83–4

quantum theory, 220, 224, 309

radio, basis of, 304

radioactivity, 87, 307, 309

railways, 274

Ramus, Pierre, 174

Regiomontanus, 153

relativity, 213, 224, 308

Renaissance, the, 133, 150, 152, 155

Roebuck, John, 267, 270

roller, the, 50

rolling-mills, 142

Rome, ancient, camps in, 23; philosophy, 85–7; at war with Carthage, 100–1; legacy of, 101–2, 117; aqueducts, 102; architecture, 119; cavalry and transport, 122; mining, 149

Römer, Olaus, 223–4

Röntgen, Wilhelm von, 305, 309

Royal Institution, 295–7, 299

Royal Observatory, Greenwich, 199–202, 217

Royal Society, the, 194, 196, 207, 222, 249, 257, 289, 291

Rudolph II, Emperor of Austria, 164

Rudolph II, Emperor of Germany, 152

Rumford, Count, see Thompson, Benjamin

Rutherford, Lord, 307, 309

safety valves, 255

Sahara Desert, 46

St Albans, Hertfordshire, 104

St Thomas Aquinas, 181

Saturn, 81, 96, 169

Savery, Captain, 258–60, 262–5

sawmills, 120–1

scale plans, 57–9

Scheiner, Christopher, 203–5

Scientific Revolution, 61, 132–3, 188, 240

scientific societies, 192–6

screw, the, 57, 59

Seleucus, 92

ships, Greek, 69

sight, 17

sledges, 50

Snell's Law, 222

Society for the Pursuit of Natural Knowledge, the, 193

Socrates, 89

solar system, 133, 134, 153–9, 160 ff.; Galileo on, 168–73, 188, 207–8, 211, 224; model of, in Jupiter, 188; dynamics of, 210–12; Newton on, 213–14; Gilbert's mistaken theories about, 281

How Things Work
The Universal Encyclopedia of Machines

How is colour television transmitted?
What makes an electric refrigerator cold?
Why is dry cleaning called dry?
How does an electric computer compute?

How Things Work supplies readily-understood answers to all
the questions that perplex the layman and puzzle the
inquisitive child in the age of the machine. It explains
everything from the simplest household gadget to the most
complex industrial process, the theory and practice of modern
machines and methods from the ball-point pen to the jet
engine, from the polaroid camera to radar.

1,071 two-colour drawings and sixteen pages of four-colour
illustrations, each facing a page of explanatory text, are
arranged according to the common principles underlying the
machines and processes discussed – optics, internal
combustion, electronics etc – providing a fascinating and
concise introduction to all the major sources of modern
technology. A full index at the back of the book lists
alphabetically all the individual items included.
Whether read in its entirety, dipped into at random, or used
simply as a straightforward reference book, *How Things Work*
is an indispensable book for everyone who has ever wondered
why a zipper zips, what makes a ship float, how a gyroscope
spins . . .

Man Modified
An Exploration of the Man/Machine Relationship

David Fishlock

The basic theme of this book is a brand-new technology, man/machine interaction. From an examination of the complex and versatile mechanism of the human body, *Man Modified* goes on to consider how man has used his ingenuity to extend his capacities: artificial kidneys, hearts and larynxes; transplants; computers; implanted heart pacemakers; artificial hip joints and the virtues of plastics; prostheses for people with amputations and for thalidomide babies; heart–lung machines; microminiaturised circuits; radio pills; exoskeletons and cyborgs.

'A timely work . . . whilst this book should be of interest to all socially responsible people, whether scientists or not, it is almost essential reading for those in or about the Life Sciences.'
NEW SCIENTIST

'Very well done and explained for the layman.'
ECONOMIST

'He casts his net so wide that few doctors are likely to be familiar with all the projects he describes, and all will find something of interest in it.'
THE LANCET

'Fascinating and exhaustively researched . . .'
FINANCIAL TIMES

'Probably unique at present in that it brings together between two covers a wide spectrum of the applications of advanced technology to assist, evaluate and modify the human body.'
SCIENCE JOURNAL